Global Perspectives on Health Geography

Series Editor

Valorie Crooks, Department of Geography, Simon Fraser University
Burnaby, Canada

Global Perspectives on Health Geography showcases cutting-edge health geography research that addresses pressing, contemporary aspects of the health-place interface. The bi-directional influence between health and place has been acknowledged for centuries, and understanding traditional and contemporary aspects of this connection is at the core of the discipline of health geography. Health geographers, for example, have: shown the complex ways in which places influence and directly impact our health; documented how and why we seek specific spaces to improve our wellbeing; and revealed how policies and practices across multiple scales affect health care delivery and receipt.

The series publishes a comprehensive portfolio of monographs and edited volumes that document the latest research in this important discipline. Proposals are accepted across a broad and ever-developing swath of topics as diverse as the discipline of health geography itself, including transnational health mobilities, experiential accounts of health and wellbeing, global-local health policies and practices, mHealth, environmental health (in)equity, theoretical approaches, and emerging spatial technologies as they relate to health and health services. Volumes in this series draw forth new methods, ways of thinking, and approaches to examining spatial and place-based aspects of health and health care across scales. They also weave together connections between health geography and other health and social science disciplines, and in doing so highlight the importance of spatial thinking.

Dr. Valorie Crooks (Simon Fraser University, crooks@sfu.ca) is the Series Editor of Global Perspectives on Health Geography. An author/editor questionnaire and book proposal form can be obtained from Publishing Editor Zachary Romano (zachary.romano@springer.com).

Andy Sharma

Current and Emerging Rural Health Issues in the Southeast United States

Springer

Andy Sharma
Bel Aire, KS, USA

ISSN 2522-8005 ISSN 2522-8013 (electronic)
Global Perspectives on Health Geography
ISBN 978-3-032-07542-0 ISBN 978-3-032-07543-7 (eBook)
https://doi.org/10.1007/978-3-032-07543-7

This Springer imprint is published by the registered company Springer Nature Switzerland AG
The registered company address is: Gewerbestrasse 11, 6330 Cham, Switzerland

Acknowledgments and Background

I would like to express my gratitude to the Carolina Population Center (CPC) at the University of North Carolina at Chapel Hill for providing comprehensive training in demography and population health. I am thankful for the scholarships that supported my work. I also acknowledge the Southern Demographic Association (SDA), especially its long-standing members, for providing a platform where I have shared my research for the past twenty years. I am grateful to the Population Association of America (PAA) for organizing the annual conference and a series of ongoing training sessions. These events have served as a welcome place where I may share my research and engage with a community of scholars. I would like to thank Amanda Sharma for her valuable contributions to this book, including her design of the cover. Her expertise in emergency management enhances the chapter on emerging health needs and disaster preparedness. I also thank Zachary Romano for engaging in the initial concept of this work. I collectively express my gratitude to Dr. Valorie Crooks, Jan Hendrickson-Smith (former training director at the Carolina Population Center), the Center for the Study of the American South at the University of North Carolina at Chapel Hill, attendees at the Health Disparities in the United States session at the American Association of Geographers 2023 Annual Meeting, the WSU resources team, and Kristine Sande (program director at the Rural Health Information Hub).

I conclude this section by sharing my motivation for writing this book. As an educator and health researcher, I believe that both individual and community health can be enhanced by providing information that is clear and easily accessible. Additionally, by recognizing the interdisciplinary aspects of health and well-being—encompassing economics, geography, history, public policy, and sociology—we can gain a deeper understanding of population health and health-care management. This perspective is not only informed by my research but also my personal experiences. These experiences have influenced my views of health and highlighted its role in overall well-being. I am thankful for the valuable insights I gained as a funded trainee through grants from the National Institute on Aging (NIA) and the National Institute of Child Health and Human Development (NICHD). These awards supported my research and I was granted a PhD in Public

Policy with a minor in Sociology from the University of North Carolina at Chapel Hill. My lengthy academic training, which also includes an MS in Mathematics from Loyola University Chicago and an MA in Economics from DePaul University, has provided me with robust quantitative skills that I apply to my analysis of health and policy. I am committed to utilizing this foundation to contribute meaningfully to the field.

Contents

Chapter 1
Introduction

Abstract The Southeastern United States, encompassing states such as Alabama, Arkansas, Georgia, Louisiana, Mississippi, and Tennessee, presents a distinctive landscape for the examination of rurality and health. This region is characterized by significant health disparities with implications for public health policy and intervention strategies. For instance, diabetes prevalence rates in several rural counties in Mississippi rank among the highest in the nation while cardiovascular disease rates in Alabama and Louisiana exceed national averages. These health challenges reflect broader systemic issues that impact rural populations, including access to quality healthcare, lifestyle-related risk behaviors, and socio-economic factors. By examining the leading causes of mortality and the spatial distribution of prevalent health issues within this region, we can better understand the challenges faced by rural communities. This study is further enriched by a demographic examination of trends from 1850 to 2023 with a projection into 2030, thereby providing a historical context that informs contemporary health initiatives. Recognizing the relationship between demographic shifts and health outcomes aids in developing effective, culturally sensitive health programs tailored to meet the needs of these populations.

Keywords Southeastern United States · Cardiovascular disease · Diabetes · Rural · Health · Alabama · Arkansas · Georgia · Louisiana · Mississippi · Tennessee

Overview

I am pleased to offer this book as a valuable resource for gaining insight into and finding effective solutions for rural health needs in the Southeast region of the United States. With a lengthy experience in researching and teaching health disparities, health economics, population health, and spatial analysis, I provide a comprehensive examination of various health challenges faced by rural communities in Alabama, Arkansas, Georgia, Mississippi, Louisiana, and Tennessee. These areas encounter significant health needs due to factors such as elevated levels of poverty,

A. Sharma, *Current and Emerging Rural Health Issues in the Southeast United States*, Global Perspectives on Health Geography,
https://doi.org/10.1007/978-3-032-07543-7_1

outdated infrastructure, and inadequate healthcare resources. To address these pressing concerns, this book applies an interdisciplinary approach. It skillfully integrates core principles from economics, geography, history, and sociology with up-to-date research on the leading causes of mortality. Furthermore, it includes a special chapter that explores the intersection of natural hazards and emerging health considerations.

The book begins by establishing a solid foundation with a definition and overview of rural health in Chap. 1. It then delves into the historical context of rural health in Chap. 2. To comprehend health disparities in the Southeast, Chap. 3 briefly examines social and racial considerations for health. Chap. 4 focuses on current health needs by highlighting prevalent conditions, such as coronary heart disease, diabetes, and obesity. Chap. 5 addresses emerging health concerns related to floods, hurricanes, severe storms, tornadoes, and other natural hazards. Chap. 6 concludes with a discussion of emerging health issues associated with extreme heat events and provides examples for Northeast Arkansas and Southeast Louisiana. This chapter also provides a concise summary accompanied by a selection of additional resources to support further exploration and knowledge-building. With its interdisciplinary perspective and comprehensive coverage of key health issues, this book is an invaluable resource for healthcare professionals, researchers, policymakers, and anyone interested in improving the health outcomes of rural communities in the Southeast United States.

Rural and Health

Rural communities contend with unique challenges in accessing and obtaining quality healthcare, leading to adverse or poor health outcomes. Before examining these issues, defining rural and health would be useful. In general terms, rural refers to areas that are located outside of urban centers. Rural areas are typically characterized by sparse population and inadequate infrastructure development. Rural communities often have a closer connection to agriculture and natural resources. However, this is becoming less of a distinguishing factor in some areas due to excessive development. For instance, an analysis of land use change in Alabama revealed an increase in urban lands and a decrease in agricultural lands from 1990 to 2018 (Adjei et al., 2023). An earlier study revealed a notable increase in the expansion of developed land in Georgia and Tennessee with projections indicating that rapid rural development in the Southeastern region is expected to continue through 2030 (White et al., 2009). According to the United States Office of Management and Budget (OMB), there is no single definition of rural areas. Instead, the OMB provides definitions based on different criteria for different purposes. The two most commonly used definitions by OMB are:

1. Urban areas: Urban areas are defined as areas with a population of at least 50,000 people.

2. Rural areas: Rural areas are defined as areas that are not classified as urban.

The United States Census Bureau also provides definitions of rural and urban areas. The Census Bureau has two main definitions for determining urban and rural areas:

1. Urbanized areas (UA): Urbanized areas are densely developed residential, commercial, and other non-residential areas with a population of 50,000 or more.
2. Urban clusters (UC): Urban clusters are areas that consist of densely developed residential, commercial, and other non-residential areas with a population of at least 2,500 but less than 50,000.

Any areas not meeting the criteria for urbanized areas or urban clusters are considered rural by the Census Bureau (HHS, 2020; Ratcliffe et al., 2016; U.S. Census Bureau, 2020). These definitions are used for various statistical and administrative purposes, including determining funding allocations and policy decisions related to rural development.

While the concept of "rural" may not have a clear definition, the concept of health is more easily defined. According to the World Health Organization (WHO), health is described as "a state of complete physical, mental, and social well-being and not merely the absence of disease or infirmity" (WHO, 1948, pg. 100). Readers should note that this definition recognizes the inclusion of mental well-being as an essential component of overall health. In addition to physical health, mental and emotional well-being are important aspects of the overall state of wellness. Mental health encompasses an individual's emotional and psychological well-being, thereby exerting influence over their behaviors and strategies for managing stress. Conditions such as anxiety and depression can significantly impact a person's functioning and overall health.

We should acknowledge that health is a multifaceted and dynamic concept that varies from person to person. It is influenced by numerous factors, including economic status, environment, genetics, geographical location, lifestyle choices, and accessibility to healthcare services. In this book, I employ a comprehensive approach to health, examining various dimensions to promote and maintain well-being. This involves advocating for equitable access to healthcare and addressing the various aspects of health to ensure individuals can achieve optimal well-being.

Why Study Rural Health?

Rural areas continue to experience several pressing health needs. Consider diabetes, which is the top ten leading cause of mortality both nationally and globally (Ahmad et al., 2024). The Southeast continues to grapple with this public health issue that has adversely impacted the health of minority populations for the past two decades. Mississippi, in particular, suffers from disproportionately higher diabetes rates with a greater burden in rural areas (Sharma, 2014; Sharma, 2023). Diabetes can have far-reaching consequences beyond its direct health implications. Individuals with diabetes are at an increased risk of developing comorbidities such as cardiovascular disease, kidney disease, and vision loss (ElSayed et al., 2024). The financial burden associated with managing diabetes is also substantial as individuals may face difficulties affording medications, monitoring devices, and regular healthcare visits (Parker et al., 2024). In Alabama, cardiovascular disease is the leading cause of mortality and ranks among the highest in the nation (CDC, 2022). This risk is compounded in rural areas due to higher tobacco use, low levels of leisure-time physical activity, poor nutrition, and excessive alcohol consumption. Furthermore, limited access to healthcare facilities in rural Alabama limits residents' ability to receive preventive care, early screenings, and treatment for cardiovascular conditions (Andrabi et al., 2024). Cardiovascular disease is also a public health concern in Arkansas and Louisiana where age-adjusted mortality rates exceed the national average (Dwyer-Lindgren et al., 2016; Mokdad et al., 2018). This condition is exacerbated by high rates of obesity throughout the study area.

The Southeast region, particularly rural areas, not only experiences a high prevalence of preventable health conditions but also faces a shortage of healthcare professionals. According to the State of the United States Health Care Workforce report, Alabama, Arkansas, and Mississippi have the lowest number of practicing physicians with rates of 186, 187, and 166 per 100,000 population, respectively (HRSA, 2024). In contrast, states such as Maine, New York, and Oregon have an average of over 270 physicians per 100,000 population. While Georgia may not face a physician shortage concern, there is a scarcity of nursing professionals (Lyons, 2023). With only 7.2 nurses per 1,000 population in 2023 and an incremental increase to 8.7 nurses per 1,000 population in 2024, Georgia ranks among the lowest in the nation (Feeney, 2024; Robinson, 2025). By comparison, the national average is approximately 9.6 nurses per 1,000 population with most Midwest states averaging 10.3 nurses per 1,000 population.

Study Area

The study area for this research includes several states in the United States, namely Alabama, Arkansas, Georgia, Louisiana, Mississippi, and Tennessee (Fig. 1.1). These states are collectively referred to as the Southeastern United States based on their geographical location and common characteristics. Typically, this study would also encompass three additional states, Florida, North Carolina, and South Carolina, which are often included when referring to the Southeastern region. However, due to the specific focus on rural health and the aim of highlighting various policy initiatives, I decided not to include these three states in the analysis.

The decision to focus the research specifically on these six states is motivated by several factors. First, these states exhibit distinctive rural health challenges. A greater burden of poverty, lack of racial diversity, and limited social capital are attributes unique to this geographic area (Erving, 2024; Peluso et al., 2024; Reynolds & Homan, 2023). These aspects require further examination, which will be addressed

Fig. 1.1 Study area. (Source: ESRI, TomTom, Garmin, FAO, NOAA, USGS, EPA, USFWS (A. Sharma))

in detail in Chap. 2. Second, the Southeastern United States has a dynamic economic setting where agriculture and manufacturing remain integral to the local gross domestic output. This means rural communities retain their significance as substantial economic contributors, benefiting from a sufficient population base, even amid the ongoing trend of population growth and migration towards urban centers. Equally important, the region's rich history and culture (e.g., Civil Rights Movement, Indigenous population) influence the social determinants of health (Bardo & Cummings, 2023; Liu & Yang, 2022). Factors such as race, ethnicity, socio-economic status, and geographic location play a role in shaping health outcomes. By narrowing the focus to Alabama, Arkansas, Georgia, Louisiana, Mississippi, and Tennessee, I can better understand the underlying factors that contribute to these issues. Each state has its own demographic characteristics, healthcare landscape, and policy initiatives that influence rural health outcomes. By examining these states individually and collectively, I can identify common patterns, differences, and key learnings that can inform policy and practice.

Demographic Attributes

This section presents an analysis of the general population characteristics for each state within the study area, spanning from the year 1850 to the latest census in 2023 with a projection for 2030. Examining the shifts in racial and ethnic attributes, as well as the development of rural and urban status, offers valuable insights in comprehending the subsequent chapters that focus on the Westward expansion and addition of new territories in the United States, developments in healthcare delivery, and social and racial considerations for healthcare. By gaining insight into the evolution of population characteristics in this region, healthcare professionals will cultivate an appreciation when implementing culturally enriched approaches in their practice and research.

Alabama

Alabama experienced substantial population growth from the late 1800s through the 1950s (Table 1.1). The population more than doubled between 1850 and 1900 with consistent growth in each subsequent decade. This expansion can primarily be attributed to industrialization in the late nineteenth century alongside increased cotton and lumber production. Although a sizable portion of the population settled in urban areas, such as Birmingham, Mobile, and Montgomery, the majority of the population, approximately 56%, remained rural throughout most of the twentieth century.

In terms of racial composition, White alone accounted for 67% of the population, Black alone constituted 32%, and Asians, Native Americans, and Hispanics

Table 1.1 Alabama population

Year	Total population	Percentage change from previous period
1850	771,623	
1860	964,201	25%
1870	996,922	3%
1880	1,262,505	27%
1890	1,513,401	20%
1900	1,828,697	21%
1910	2,138,093	17%
1920	2,348,174	10%
1930	2,646,248	13%
1940	2,832,961	7%
1950	3,061,743	8%
1960	3,266,740	7%
1970	3,444,165	5%
1980	3,893,888	13%
1990	4,040,587	4%
2000	4,447,100	10%
2010	4,779,736	7%
2020	5,024,279	5%
2023	5,108,468	2%
2030	5,213,942	*

Source: Historical Population Change Data and the Population Estimates Program from the United States Census and *National Population Projections from the Weldon Cooper Center (2024)

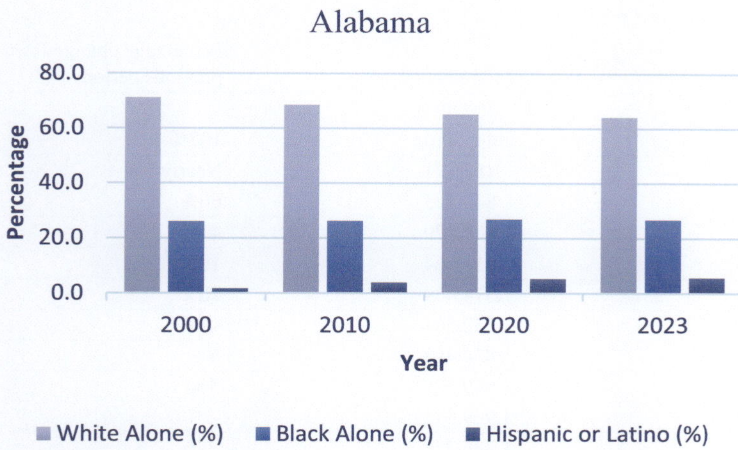

Fig. 1.2 Race and ethnicity by year

collectively made up less than 2% in 1950. This shift in demographic distribution marked a departure from previous eras where White alone comprised 55% of the population and Black alone nearly 45% (see Fig. 1.2). During the 1990s, there was a decrease in the percentage of the Black population while there was an increase in the share of the Asian and Hispanic population (U.S. Census Bureau, 2021a). As of 2025, the population exceeded five million people with 64% categorized as White alone, 27% as Black alone, 5% as Hispanic or Latino origin, and less than 2% as Asian. While the majority of the population, approximately 59%, now resides in urban areas, a sizable portion is present in rural areas.

Arkansas

Arkansas also experienced significant population growth from the late 1800s through the 1930s (Table 1.2). Unlike Alabama, the population declined from 1940 to 1960 before increasing steadily in each subsequent decade. This growth can be attributed to several factors, including the expansion of industries such as agriculture, manufacturing, and timber during the late nineteenth century. While urban areas such as Little Rock, Fort Smith, and Fayetteville saw a considerable influx of settlers, the majority of Arkansans, over 60%, continued to reside in rural areas throughout most of the twentieth century. This was due to the state's strong agricultural base and the prevalence of small farming communities.

In terms of racial composition, White alone accounted for 68% of Arkansas' population while Black alone approximated 32% in 1950. Other racial and ethnic groups, including Asians, Hispanics, and Native Americans, collectively constituted less than 1% of the population. This marked a difference from previous eras where White alone comprised a lower percentage of the population and Black alone constituted a larger portion (see Fig. 1.3).

Table 1.2 Arkansas population

Year	Total population	Percentage change from previous period
1850	209,897	
1860	435,450	107%
1870	484,471	Nearly 0%
1880	802,525	66%
1890	1,128,179	41%
1900	1,311,564	16%
1910	1,574,449	20%
1920	1,752,204	11%
1930	1,854,482	6%
1940	1,949,387	5%
1950	1,909,511	-2%
1960	1,786,272	-6%
1970	1,923,295	8%
1980	2,286,435	19%
1990	2,350,725	3%
2000	2,673,400	14%
2010	2,915,918	9%
2020	3,011,524	3%
2023	3,067,732	2%
2030	3,084,796	*

Source: Historical Population Change Data and the Population Estimates Program from the United States Census and *National Population Projections from the Weldon Cooper Center (2024)

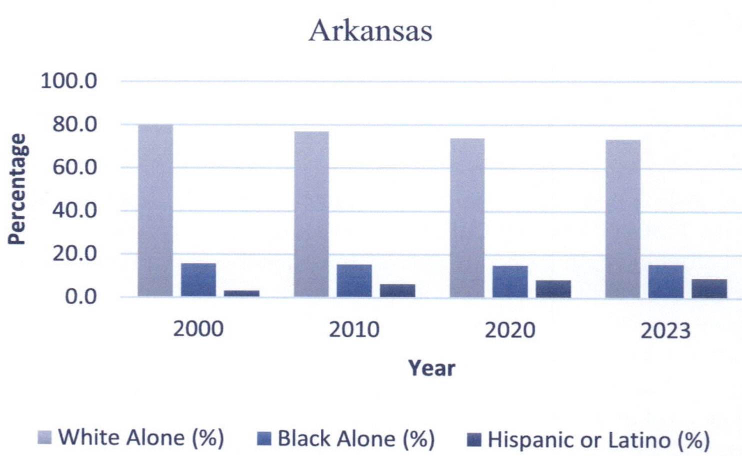

Fig. 1.3 Race and ethnicity by year

During this century, Arkansas witnessed shifts in its demographic distribution (U.S. Census Bureau, 2021b). There was an increase in the Hispanic and Latino population. According to Census estimates for the year 2025, the population of Arkansas has exceeded three million people with 70% categorized as White alone, 15% as Black alone, 9% as Hispanic or Latino origin, and less than 2% as Asian. While 60% of Arkansans now reside in urban areas, including cities such as Fayetteville, Little Rock, and Springdale, a considerable portion still lives in rural regions, highlighting the continued importance of agriculture (e.g., cotton, rice, soybeans) and rural communities in the state (Miller & Wheeler, 2021).

Georgia

Georgia, similar to other states in the southern region of the United States, has also experienced population growth (Table 1.3). According to Census estimates for the year 2025, the population of Georgia has surpassed 11 million people. This growth can be attributed to several factors, including a strong economy and a favorable business environment.

In terms of racial composition, the percentage of the population that is White alone has decreased from earlier periods but still constitutes the majority and accounts for approximately 51%. The Black alone population represents 32% of the population, reflecting a significant African American presence in the state (see Fig. 1.4). The Hispanic or Latino population has seen a steady increase and now constitutes 11% of the population. Asians approximate 5% of the population, displaying the growing influence of the Asian community. The distribution of Georgia's population also reflects shifting trends. Urban areas in Georgia, such as Atlanta, Augusta, and Savannah, have experienced substantial growth and continue to attract residents. Approximately 80% of Georgians now reside in urban areas, showing a significant increase in urbanization. This trend can be attributed to the previously mentioned expanding job market, cultural amenities, and quality of life that urban centers offer. However, rural areas still play a role in the state (Tanner, 2021). In particular, the agriculture and natural resource industry (e.g., cotton, peanuts, timber) generates substantial revenue for the state with an estimated amount of over $12 billion (Kane, 2022). Another demographic attribute to note is the percentage of the population that is aged 65 and older has increased in rural areas and will continue to do so until 2030. This presents various rural healthcare challenges for Georgia, which will be discussed in Chap. 4.

The changing demographics and population distribution in Georgia reflect the ongoing transformation of the state (U.S. Census Bureau, 2021c). The growth and changing population characteristics have contributed to the state's economic development and cultural vibrancy. Georgia's status as a hub for industries related to aerospace, automotive, logistics, and manufacturing has attracted individuals from various backgrounds, resulting in a diverse and dynamic population.

Table 1.3 Georgia population

Year	Total population	Percentage change from previous period
1850	906,185	
1860	1,057,286	17%
1870	1,184,109	12%
1880	1,542,180	30%
1890	1,837,353	19%
1900	2,216,331	21%
1910	2,609,121	18%
1920	2,895,832	11%
1930	2,908,506	Nearly 0%
1940	3,123,723	7%
1950	3,444,578	10%
1960	3,943,116	14%
1970	4,589,575	16%
1980	5,463,105	19%
1990	6,478,216	19%
2000	8,186,453	26%
2010	9,687,653	18%
2020	10,711,908	11%
2023	11,029,227	3%
2030	11,534,245	*

Source: Historical Population Change Data and the Population Estimates Program from the United States Census and *National Population Projections from the Weldon Cooper Center (2024)

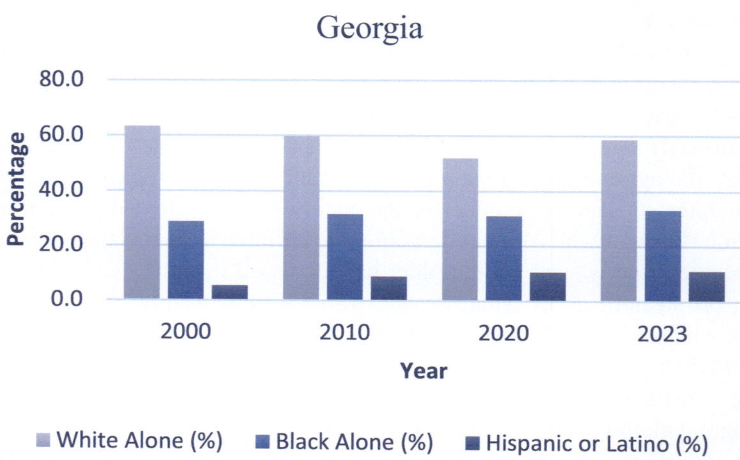

Fig. 1.4 Race and ethnicity by year

Louisiana

Unlike other states in the Southeast, Louisiana has experienced inconsistent population dynamics (Table 1.4). More specifically, the population increased significantly from 1950 to 1980 then experienced minimal gains until 2010. After 2016, the population declined. According to Census estimates for the year 2025, the population of Louisiana was approximately 4.7 million people. This minimal growth can be attributed to a range of interacting factors with out-migration serving as a contributor. Key elements of this trend include economic considerations related to unemployment, lack of public investment in education, and relocation to be closer to family.

Louisiana may have an increasingly diverse population but the majority remains White alone at 57%. Black alone represents 32% of the population, highlighting the state's history and African American influence. Approximately 6% of the population is Hispanic or Latino. Asians represent a small percentage and only constitute 2% of the population. The population distribution in Louisiana has also changed over time. While cities such as Baton Rouge, New Orleans, and Shreveport are major population centers, there has been significant growth in suburban areas around these cities (e.g., suburban parishes near Baton Rouge and the Texas border). These suburbs offer a mix of urban amenities and a quieter lifestyle, attracting residents seeking a balance. However, rural areas still play a role, particularly in the agriculture (e.g., rice, soybeans, sugarcane) and forestry industry (Guidry et al., 2023). While the percentage of the population in rural areas has declined during the last century, nearly 26% still reside in this region.

The slowly changing demographics in Louisiana reflect the state's ongoing development (U.S. Census Bureau, 2021d; see Fig. 1.5). Louisiana's distinctive cultural heritage, characterized by its Cajun and Creole culinary traditions, prominent festivals such as Mardi Gras, and rich musical legacy, continues to attract individuals from a variety of cultural and ethnic backgrounds. However, the population may experience an ongoing decline over the next decade due to out-migration to Alabama, Arkansas, Tennessee, and Texas.

Table 1.4 Louisiana population

Year	Total population	Percentage change from previous period
1850	517,762	
1860	708,002	37%
1870	726,915	3%
1880	939,946	29%
1890	1,118,587	19%
1900	1,381,625	24%
1910	1,656,388	20%
1920	1,798,509	9%
1930	2,101,593	17%
1940	2,363,880	12%
1950	2,683,516	14%
1960	3,257,022	21%
1970	3,641,306	12%
1980	4,205,900	16%
1990	4,219,973	Nearly 0%
2000	4,468,976	6%
2010	4,533,372	1%
2020	4,657,757	3%
2023	4,573,749	-2%
2030	4,752,752	*

Source: Historical Population Change Data and the Population Estimates Program from the United States Census and *National Population Projections from the Weldon Cooper Center (2024)

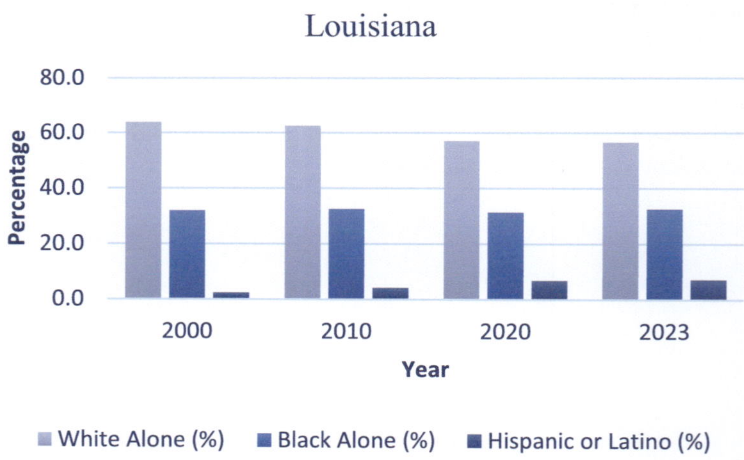

Fig. 1.5 Race and ethnicity by year

Mississippi

Mississippi experienced steady population growth until 1940, albeit at a slower pace compared to other states in the southern region (Table 1.5). From 1940 to 1970, the population was stable before slowly increasing until 2016. Since then, the population has decreased each year (U.S. Census Bureau, 2021e). Census estimates show that the population of Mississippi was approximately 2.9 million people in 2025. Mississippi has a predominantly White alone population at 56%. Black alone constitutes the largest minority group, comprising 37% of the population. The state has a small Hispanic or Latino population, accounting for 3% of the residents (see Fig. 1.6). The remaining population is composed of various other racial and ethnic groups, including Asians and Native Americans.

The population distribution in Mississippi is characterized by a mix of rural and urban areas. Cities such as Jackson, the state capital, Gulfport, and Hattiesburg serve as major population centers. However, rural areas are important to the state's economy, particularly in agriculture (e.g., corn, soybeans) and forestry (Coblentz, 2022). Nearly 50% of the population still resides in rural areas. Mississippi's population is influenced by a range of factors. The state's low cost of living, as well as options for affordable housing, continues to appeal to residents. Mississippi's rich cultural heritage, including its unique music, literature, and culinary traditions, also contributes to its charm. The state's history, particularly its role in the Civil Rights Movement (1954–1968), has left a lasting impact on its population and continues to shape its identity. Mississippi's efforts to diversify its economy have led to job opportunities in industries such as manufacturing and green energy production (Harrison, 2024). Another recent development corresponds to investments in internet technology and data storage capabilities. While Mississippi's population growth has been slower compared to other states in the region during the past century and may continue to decrease for the upcoming decade, the state's efforts to attract new residents and promote economic development will have implications for rural communities. The state's investment in urban and non-urban counties outside Jackson and Southaven may improve rural healthcare access. Specifically, improvements in infrastructure, the expansion of business services, and the development of connectivity between rural and urban regions create possibilities for greater access to healthcare. Enhanced transportation networks, characterized by improved road conditions and greater geographic coverage, may reduce travel expenses and decrease transit times, thereby lowering barriers to obtaining care.

Table 1.5 Mississippi population

Year	Total population	Percentage change from previous period
1850	606,526	
1860	791,805	31%
1870	827,922	5%
1880	1,131,597	37%
1890	1,289,600	14%
1900	1,551,270	20%
1910	1,797,114	16%
1920	1,790,618	0%
1930	2,009,821	12%
1940	2,183,796	9%
1950	2,178,914	Nearly 0%
1960	2,178,141	Nearly 0%
1970	2,216,912	2%
1980	2,520,638	14%
1990	2,573,216	2%
2000	2,844,658	11%
2010	2,967,297	4%
2020	2,961,279	Nearly 0%
2023	2,939,690	-1%
2030	2,956,772	*

Source: Historical Population Change Data and the Population Estimates Program from the United States Census and *National Population Projections from the Weldon Cooper Center (2024)

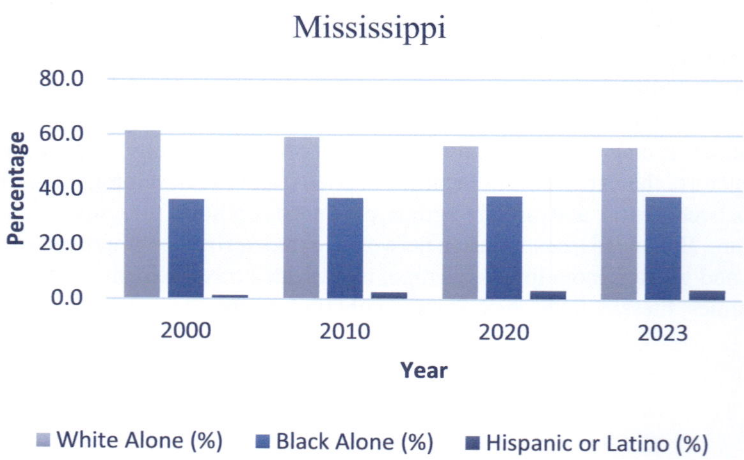

Fig. 1.6 Race and ethnicity by year

Tennessee

Tennessee has experienced consistent population growth in recent years with approximately 7.1 million people residing in the state based on Census estimates for 2025 (Table 1.6). This growth has outpaced many neighboring states in the southern region. The population of Tennessee is characterized by racial and ethnic diversity with multiple groups contributing to its demographic composition. The largest group is still White alone, comprising around 73% of the population. Black alone constitutes the second-largest group, accounting for approximately 16% of residents. The state also has a growing Hispanic or Latino population, representing about 6% of residents (see Fig. 1.7). The remaining portion of the population is composed of other minority groups such as Asians and Native Americans.

The demographic characteristics of Tennessee demonstrate a diverse combination of urban and rural regions. Notably, urban centers such as Memphis, Nashville, and Knoxville function as population hubs, granting access to various employment prospects and enriching cultural experiences. However, nearly 32% of the population resides in rural areas, thereby maintaining a substantial influence on the state's economy. More specifically, agriculture (e.g., corn, soybeans) and the production of tobacco generate substantial revenue.

Table 1.6 Tennessee population

Year	Total population	Percentage change from previous period
1850	1,002,717	
1860	1,109,801	11%
1870	1,258,520	13%
1880	1,542,359	23%
1890	1,767,518	15%
1900	2,020,616	14%
1910	2,184,789	8%
1920	2,337,885	7%
1930	2,616,556	12%
1940	2,915,841	11%
1950	3,291,718	13%
1960	3,567,089	8%
1970	3,923,687	10%
1980	4,591,120	17%
1990	4,877,185	6%
2000	5,689,283	17%
2010	6,346,105	12%
2020	6,910,840	9%
2023	7,126,489	3%
2030	7,359,519	*

Source: Historical Population Change Data and the Population Estimates Program from the United States Census and *National Population Projections from the Weldon Cooper Center (2024)

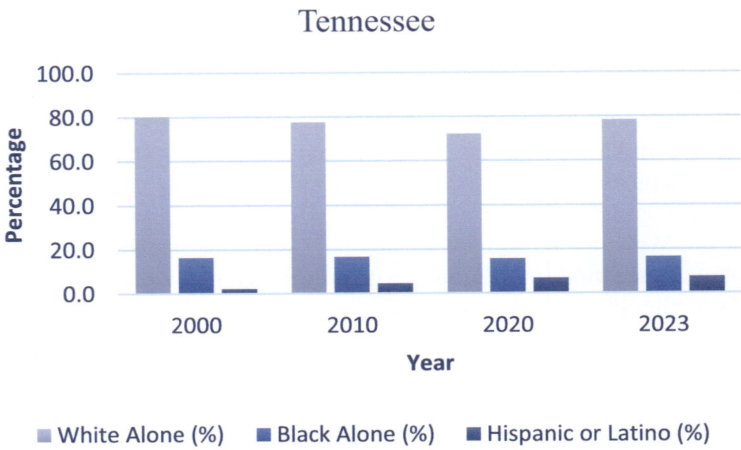

Fig. 1.7 Race and ethnicity by year

Tennessee's population growth can be attributed to a variety of factors (U.S. Census Bureau, 2021f). The state's lower cost of living and diverse economy appeal to individuals and families. The manufacturing sector and tourism also provide employment options. Tennessee features a prominent musical legacy with a particular emphasis on genres including blues, country, and rock 'n' roll. The state's vibrant music scene, coupled with its historical sites and picturesque landscapes, establishes it as an attractive destination for tourists and new residents. Overall, Tennessee's population will continue to grow for the upcoming decade.

Demographic Outlook

The Southeastern United States is notable for having the highest proportion of individuals identifying as Black alone in the nation. This demographic trend is particularly evident in specific counties within Alabama and Mississippi, where the reported estimates exceed 80%. This pattern is expected to continue throughout this decade, reflecting the enduring historical and socio-cultural dynamics that shape the demographics of the area. An examination of the overall population characteristics in the study area, as depicted in Figs. 1.2, 1.3, 1.4, 1.5, 1.6, and 1.7, indicates that the Black alone population has remained relatively stable while the White alone population has experienced a decline from 2000 to 2020. In contrast, the Hispanic population has seen substantial growth and is projected to continue increasing over the next decade. Currently, Hispanics represent the fastest-growing demographic group, accounting for 19% of the overall national population. Census projections suggest this figure will increase to 21% by 2030. While the Hispanic population remains predominantly concentrated in states such as California, Texas, Florida,

New York, and Arizona, recent demographic trends indicate growth in Alabama, Arkansas, Georgia, and Tennessee.

Readers will be interested to note that in March 2024, the Office of Management and Budget (OMB) updated its standards for the maintenance, collection, and presentation of race and ethnicity data applicable to all federal agencies for the first time in 27 years. The new Statistical Policy Directive No. 15 (SPD 15) introduces a combined category for race and ethnicity and adds a distinct category for individuals identifying as Middle Eastern or North African. Under this directive, separate inquiries regarding race and Hispanic ethnicity have been eliminated. As a result, race and ethnicity will now be considered as a single category. Additionally, the directive removes outdated terminology. For more information, refer to the Federal Register (https://www.federalregister.gov/documents/2024/03/29/2024-06469/revisions-to-ombs-statistical-policy-directive-no-15-standards-for-maintaining-collecting-and) or access the final version (https://spd15revision.gov/). These updates aim to improve the accuracy and relevance of data collection across federal agencies.

Concluding Remarks

This book focuses on the challenges of rural health in the Southeastern United States, specifically in Alabama, Arkansas, Georgia, Louisiana, Mississippi, and Tennessee. It examines the distinctive health issues faced by rural communities due to limited healthcare infrastructure, shortages of medical professionals, and socioeconomic factors. By focusing on a specific region, this book aims to provide a comprehensive analysis of rural health issues and inform practical solutions and interventions for improving health outcomes in the Southeastern United States.

In addition to discussing the challenges faced by rural communities in the Southeastern United States, this book delves into the underlying factors contributing to these health outcomes. It examines historical, social, economic, and policy contexts that have contributed to the current state of rural health in the region. By contextualizing the issues, this book aims to provide an inclusive understanding of the complex factors that impact health in these communities.

This introductory chapter also provides an overview of the demographic characteristics of Alabama, Arkansas, Georgia, Louisiana, Mississippi, and Tennessee. The population growth in these states can be attributed to several factors, such as industrialization, economic development, and cultural attractions. The demographic characteristic of these states has also evolved over time with changes in the percentages of Asian, Black, Hispanic, White, and other racial/ethnic groups. While urban areas have experienced significant growth, rural areas continue to play a role in the economy, particularly in the agriculture and natural resource industries. The changing demographics and population distribution in these states present challenges and opportunities for healthcare and policy considerations (Sharma, 2016).

Overall, this book offers a comprehensive examination of rural health in the Southeastern United States, combining academic research, real-world examples, and community perspectives (O'Connell et al., 2023). It seeks to raise awareness of the pressing health issues faced by rural communities in the region and inspire action to address these challenges. By providing insights into the unique context of the Southeast, this book holds the potential to inform evidence-based and culturally sensitive interventions that can contribute to improved health equity in rural areas.

Appendix

In academic discourse about cartography, the use of technical terminology is prevalent, as it allows for communication of complex ideas. To enhance reader understanding, key terms will be defined systematically. References and links to additional resources will also be provided for those seeking further insight into the concepts. This approach aims to bridge the gap between specialized language and accessibility, fostering a more inclusive dialogue surrounding cartographic representation and its implications for health and policy.

Types of Maps

A choropleth map is a thematic map that employs color or shading to represent data values for specific geographic regions. These regions may be countries, regions, states, counties, cities, blocks, or any other unit (referred to as a polygon in general terms). These maps are utilized to visually depict spatial patterns and variations in data across different areas. In a choropleth map, each region or area is color-filled or shaded based on its individual value or range of values. The intensity of the color or shading indicates the relative magnitude of the variable being measured. In most cases, darker colors signify higher values while lighter colors represent lower values. Choropleth maps find extensive application in the display of statistical data, encompassing population characteristics (such as age, gender, and race), economic attributes (such as income, poverty, and unemployment), or any other variable that can be aggregated and presented for specific geographic areas or boundaries.

A predominant category map, within the environment of Esri software, is a valuable tool for exploring and analyzing the distribution of a categorical variable across a specific geographic area. By visually representing the prevalent category or class in each location of the map, this type of thematic map facilitates the identification of spatial patterns. For example, if the variable being examined is "Race" and the designated categories are "White alone," "Black alone," and "Other race," the predominant category map highlights which race type is most commonly found in each geographic region. Diverse colors are typically employed to visually differentiate the categories with emphasis placed on the dominant category. Overall, this map provides valuable insights into the concentration of a particular category within different areas, aiding in the understanding of spatial distribution.

Another useful map refers to hot spot analysis. While red and blue colors or red and blue shading are typically used for both hot spot analysis and heat maps, these are two different maps. A heat map is used to reference (i.e., visualize specific areas) for greater exploration. In contrast, hot spot analysis utilizes statistical techniques (e.g., Getis-Ord Gi*) to calculate the extent of clustering or dispersion for a selected variable. By analyzing the spatial relationship between neighboring features, the tool identifies areas with high values surrounded by other high values (hot spots) or areas with low values surrounded by other low values (cold spots). The hot spot map displays the results of this analysis by using colors to represent areas of high and low statistical significance (Mitchell, 2020). In most cases, hot spots are represented with darker colors (e.g., red) and cold spots with lighter colors (e.g., blue). This map allows users to visualize and interpret patterns of concentration or dispersion, thereby identifying areas of interest for greater research or interventions.

Spatial Software and Datasets

With the continuous progress in spatial software and computing capabilities, researchers now have access to a range of mapping tools. One widely utilized tool is ArcGIS, which is developed by the private company Esri (Environmental Systems Research Institute). For readers unfamiliar with the term, GIS stands for Geographic Information System. According to the U.S. Geological Survey (USGS), the system can be used to analyze and display information that includes data associated with specific geographic locations. Esri also offers a suite of other mapping tools collectively known as Esri products, including ArcGIS Online, ArcGIS Pro, and location data. These products cater to the needs of individuals working with spatial data. However, the acquisition cost of Esri software can be substantial.

In contrast, QGIS (also referred to as Quantum Geographical Information Systems) serves as a free and open-source alternative for spatial data analysis (see https://qgis.org/). QGIS benefits from the contributions of a community of volunteers who support its development and maintenance. This GIS software has obtained widespread adoption among researchers, professionals, and individuals with an interest in spatial analysis. A notable advantage of QGIS is its extensibility, enabling users to enhance its functionality through plugins and scripting. This open-source attribute empowers users to customize the software to suit their specific requirements. Moreover, QGIS is compatible with multiple operating systems, including Linux, macOS, and Windows. This versatility makes it a practical option for GIS professionals who work on different platforms.

Multiple datasets were utilized in the creation of the maps in QGIS and ArcGIS Online. Census data, widely recognized for its relevance in demographic research, was employed with the utilization of the American Community Survey (ACS) five-year data from 2019 to 2023 to develop thematic maps depicting race and ethnicity (see https://www.census.gov/programs-surveys/acs/). The 2020 Decennial Census was used to develop the rural and urban map in Chap. 2 (refer to https://www.census.gov/programs-surveys/decennial-census/decade/2020/2020-census-main.html). The Centers for Disease Control and Prevention's Places: Local Data for Better Health for 2023 was utilized to map health measures such as coronary heart disease, diabetes, and obesity in Chap. 4 (see https://www.cdc.gov/places/index.html). Lastly, the Health Resources and Services Administration (HRSA) maintains comprehensive data on geographic areas designated as Health Professional Shortage Areas or HPSA (see https://data.hrsa.gov). I provide a map for the HPSA primary care scores for 2023 for the study area.

Concepts

As a result of my interdisciplinary approach to addressing rural health issues in the Southeast United States, readers will benefit by reviewing the following key concepts: race and ethnicity, social determinants of health, and health systems. I also encourage a deeper examination of these topics through supplementary readings, which can enhance both current understanding and future professional endeavors. Engaging with this material will provide valuable insights into the complexities of rural health and its broader impact on communities.

Race and ethnicity: Race typically refers to the classification of people based on physical characteristics, such as facial features or skin color. Race can influence a person's identity and social experiences, often shaping their relations and interactions in society. It can be viewed as a social construct (i.e., defined by society) or a strict biological category (i.e., genetics). Common racial categories include Asian, Black, White, Indigenous, among others, and these classifications can differ across regions and over time. For instance, the United States Census Bureau defines race and ethnicity based on self-reported responses that align with the 1997 Office of Management and Budget (OMB) standards:

> White—A person having origins in any of the original peoples of Europe, the Middle East, or North Africa.
>
> Black or African American—A person having origins in any of the Black racial groups of Africa.
>
> American Indian or Alaska Native—A person having origins in any of the original peoples of North and South America (including Central America) and who maintains tribal affiliation or community attachment.
>
> Asian—A person having origins in any of the original peoples of the Far East, Southeast Asia, or the Indian subcontinent including, for example, Cambodia, China, India, Japan, Korea, Malaysia, Pakistan, the Philippine Islands, Thailand, and Vietnam.
>
> Native Hawaiian or Other Pacific Islander—A person having origins in any of the original peoples of Hawaii, Guam, Samoa, or other Pacific Islands.

Since 2000, the census has allowed individuals to identify with more than one race (U.S. Census, 2022). As of March 2024, OMB updated the standards for maintaining, collecting, and presenting race and ethnicity data as outlined in Statistical Policy Directive No. 15 (SPD 15). This revision combines race and ethnicity into a single category, and a new category has been introduced for individuals identifying as Middle Eastern or North African (see below for updated categories). All federal agencies must implement these changes by March 2029.

In contrast, ethnicity pertains to the shared cultural, historical, and social traditions of a group and not physical characteristics. Ethnicity encompasses aspects such as language, religion, traditions, and national origin (Fenton, 2010). Examples of ethnic groups include Hispanic or Latino, Irish, Italian, and many others. Ethnicity is multifaceted and individuals may identify with several ethnicities. A shared ethnicity often arises from common cultural expressions, rituals, and community connections that foster a sense of belonging. While race and ethnicity represent different frameworks, they often intersect. For instance, an individual may identify as Black (race) while also embracing Afro-Latino (ethnicity), illustrating a blend of physical and cultural identities.

Updated categories for race and ethnicity from the 2024 OMB standards:

White—Individuals with origins in any of the original peoples of Europe, including, for example, English, German, Irish, Italian, Polish, and Scottish.

Black or African American—Individuals with origins in any of the Black racial groups of Africa, including, for example, African American, Jamaican, Haitian, Nigerian, Ethiopian, and Somali.

American Indian or Alaska Native—Individuals with origins in any of the original peoples of North, Central, and South America, including, for example, Navajo Nation, Blackfeet Tribe of the Blackfeet Indian Reservation of Montana, Native Village of Barrow Inupiat Traditional Government, Nome Eskimo Community, Aztec, and Maya.

Asian—Individuals with origins in any of the original peoples of Central or East Asia, Southeast Asia, or South Asia, including, for example, Chinese, Asian Indian, Filipino, Vietnamese, Korean, and Japanese.

Native Hawaiian or Pacific Islander—Individuals with origins in any of the original peoples of Hawaii, Guam, Samoa, or other Pacific Islands, including, for example, Native Hawaiian, Samoan, Chamorro, Tongan, Fijian, and Marshallese.

Middle Eastern or Northern Africa—Individuals with origins in any of the original peoples of the Middle East or North Africa, including, for example, Lebanese, Iranian, Egyptian, Syrian, Iraqi, and Israeli.

Hispanic or Latino—Includes individuals of Mexican, Puerto Rican, Salvadoran, Cuban, Dominican, Guatemalan, and other Central or South American or Spanish culture or origin.

Social determinants of health: Social determinants of health (SDOH) refer to the "conditions in which people are born, grow, live, work, and age and the wider set of forces and systems shaping the conditions of daily life" (WHO, n.d.). These determinants encompass a wide range of social, economic, and environmental factors, extending beyond access to healthcare (HHS, n.d.). Key categories include:

Economic stability: Various determinants such as income, employment status, and the economic environment can influence health outcomes. For instance, lower income levels may restrict access to healthcare services, nutritious food, and safe living conditions, thereby adversely affecting overall health and well-being.

Education: Educational attainment and literacy influence an individual's ability to understand health information and participate in health-promoting behaviors. A large body of research indicates that higher levels of education are positively associated with better health outcomes. Individuals with advanced education can more easily understand complex health-related information and make informed decisions regarding their health.

Healthcare access: Access to healthcare services and quality of care received are vital for maintaining health. Barriers to accessing these services can result in disparities in health outcomes.

Neighborhood and built environment: The physical surroundings, including housing quality and neighborhood safety, play a role in health outcomes. Residing in supportive and secure neighborhoods is associated with the promotion of healthier behavioral practices. Communities characterized by well-maintained infrastructure and accessible neighborhood parks are often perceived as a safe environment. This perception can increase residents' participation in outdoor activities, such as exercising, running, or walking.

Social and community context: The interplay between social networks, interpersonal relationships, and active community engagement can influence mental health outcomes. Research indicates that positive social connections can confer protective health benefits, thereby contributing to overall psychological well-being.

Understanding these determinants is essential for policymakers, healthcare providers, and communities as they design and implement effective strategies aimed at reducing health disparities and improving the overall health of the population. Another important concept to recognize is health systems. The Agency for Healthcare Research and Quality (AHRQ) characterizes a health system as "an organization that comprises at least one hospital and at least one group of physicians, providing comprehensive care—including both primary and specialty services—where these entities are linked through common ownership or joint management" (AHRQ, 2023). Health systems facilitate access to care, support medical research, promote health equity, and contribute to overall population health and well-being. They also play an important role in responding to public health emergencies and natural hazards, which are the topics that will be examined in Chap. 5.

References

Adjei, E., Li, W., Narine, L., & Zhang, Y. (2023). What drives land use change in the southern US? A case study of Alabama. *Forests, 14*(2), 171.

Agency for Healthcare Research and Quality, AHRQ. (2023). *Defining Health Systems*. Agency for Healthcare Research and Quality, AHRQ. https://www.ahrq.gov/chsp/defining-health-systems/index.html

Ahmad, F. B., Cisewski, J. A., & Anderson, R. N. (2024). *Leading causes of death in the US, 2019–2023* (Vol. 332, p. 957). JAMA.

Andrabi, M., Key, B., Francisco, M., Jeremiah, M., Mumba, M., Radhakrishnan, K., & Ball, K. (2024). A quantitative study to assess cardiovascular risks among older African American women living in rural black belt areas of Alabama. *Public Health Nursing, 41*, 979.

Bardo, A. R., & Cummings, J. L. (2023). Life, longevity, and the pursuit of happiness: The role of disability in shaping racial and sex disparities in living a long and happy life. *Population Research and Policy Review, 42*(4), 72.

Centers for Disease Control and Prevention, National Center for Health Statistics, CDC. (2022). *Heart disease mortality by state*. CDC. https://www.cdc.gov/nchs/pressroom/sosmap/heart_disease_mortality/heart_disease.htm

Coblentz, B. (2022). *Mississippi agriculture hits $9.7B overall value*. Mississippi State University Extension Service. https://extension.msstate.edu/news/feature-story/2022/mississippi-agriculture-hits-97b-overall-value.

Dwyer-Lindgren, L., Bertozzi-Villa, A., Stubbs, R. W., Morozoff, C., Kutz, M. J., Huynh, C., & Murray, C. J. (2016). US county-level trends in mortality rates for major causes of death, 1980-2014. *JAMA, 316*(22), 2385–2401.

ElSayed, N. A., Aleppo, G., Aroda, V. R., Bannuru, R. R., Brown, F. M., Bruemmer, D., & Gabbay, R. A. (2024). 10. Cardiovascular disease and risk management: Standards of care in diabetes—2023. *Diabetes Care, 47*, S158.

Erving, C. L. (2024). Stress exposure and cardiovascular disease risk among US black women: Ethnicity and nativity intersections. *Population Research and Policy Review, 43*(3), 36.

Feeney, A. (2024). *The U.S.* Nursing shortage: A state-by-state breakdown. https://nursejournal.org/articles/the-us-nursing-shortage-state-by-state-breakdown/

Fenton, S. (2010). *Ethnicity*. Polity Press.

Guidry, K., Goyal, R., & Gould, F. H. (2023). *Highlights of Louisiana agriculture 2023*. Louisiana State University. https://www.lsuagcenter.com/articles/page1720800061063

Harrison, B. (2024). Lawmakers pass $350 million deal to lure major green energy plant to north Mississippi. *Mississippi Today* (2024, January 18). https://mississippitoday.org/2024/01/18/special-session-economic-development-passes/

Health Resources and Service Administration, HRSA. (2024). *State of the U.S. Health Care Workforce, 2023. (2024 May)*. U.S. Department of Health & Human Services. https://bhw.hrsa.gov/sites/default/files/bureau-health-workforce/data-research/state-of-the-health-workforce-report-2023.pdf

Kane, S. (2022). *AgSnapshots 2022*. University of Georgia. https://caed.uga.edu/content/dam/caes-subsite/caed/publications/ag-snapshots/2022CAEDAgSnapshotsWeb.pdf

Liu, H., & Yang, T. C. (2022). Examining the reciprocity between perceived discrimination and health: A longitudinal perspective. *Population Research and Policy Review, 41*(4), 1757–1777.

Lyons, J. (2023). *Nursing shortage MAY be worse by state than projected, new analysis shows*. https://nurse.org/education/nursing-shortage-by-state-analysis/

Miller, W., & Wheeler, E. (2021). *Rural profile of Arkansas: Social & Economic Trends Affecting Rural Arkansas*. University of Arkansas, Division of Agriculture. https://www.uaex.uada.edu/publications/pdf/MP564.pdf

Mitchell, A. (2020). The ESRI® guide to GIS analysis. In *Geographic patterns and relationships* (Vol. 1). Environmental Systems Research Institute.

Mokdad, A. H., Ballestros, K., Echko, M., Glenn, S., Olsen, H. E., Mullany, E., & US Burden of Disease Collaborators. (2018). The state of US health, 1990-2016: Burden of diseases, injuries, and risk factors among US states. *JAMA, 319*(14), 1444–1472.

O'Connell, S., Boles, J., McClellan, R., & Demers, D. (2023). Mapping food security in Arkansas. *Applied Geography, 158*, 103020.

Parker, E. D., Lin, J., Mahoney, T., Ume, N., Yang, G., Gabbay, R. A., ElSayed, N. A., & Bannuru, R. R. (2024). Economic costs of diabetes in the US in 2022. *Diabetes Care, 47*(1), 26–43.

Peluso, A., Tuccillo, J., Sparks, K., Kapadia, A., & Hanson, H. A. (2024). Spatial analysis of social capital and community heterogeneity at the United States county level. *Applied Geography, 162*, 103168.

Ratcliffe, M., Burd, C., Holder, K., & Fields, A. (2016). Defining rural at the US Census Bureau. *American Community Survey and Geography Brief, 1*(8), 1–8.

Reynolds, M. M., & Homan, P. A. (2023). Income support policy packages and birth outcomes in US states: An ecological analysis. *Population Research and Policy Review, 42*(4), 73.

Robinson, C. (2025). *The U.S.* Nursing shortage: A state-by-state breakdown. https://nursejournal.org/articles/the-us-nursing-shortage-state-by-state-breakdown/

Sharma, A. (2014). Spatial analysis of disparities in LDL-C testing for older diabetic adults: A socio-environmental framework focusing on race, poverty, and health access in Mississippi. *Applied Geography, 55*, 248–256.

Sharma, A. (2016). Migration of older adults. In A. Farazmand (Ed.), *Global encyclopedia of public administration, public policy, and governance*. Springer.

Sharma, A. (2023). Exploratory spatial analysis of food insecurity and diabetes: An application of multiscale geographically weighted regression. *Annals of GIS, 29*(4), 485–498.

Tanner, D. (2021). *Rural Georgia in focus*. University of Georgia. https://www.house.ga.gov/Documents/CommitteeDocuments/2021/Rural_Development_Council/CVIOG%202020%20Census%20Data.pdf

U.S. Census Bureau. (2020). *Understanding and using American community survey data: What users of data for rural areas need to know*. U.S. Department of Commerce, Economics and Statistics Administration, U.S. Census Bureau. https://www.census.gov/content/dam/Census/library/publications/2020/acs/acs_rural_handbook_2020_ch01.pdf

U.S. Census Bureau. (2022). *About the topic of race*. U.S. Department of Commerce. https://www.census.gov/topics/population/race/about.html

U.S. Census Bureau, & America Counts Staff. (2021a). *Alabama population grew 5.1% since 2010, Surpassing 5 million*. U.S. Department of Commerce. https://www.census.gov/library/stories/state-by-state/alabama-population-change-between-census-decade.html

U.S. Census Bureau, & America Counts Staff. (2021b). *Arkansas population topped 3 million in 2020*. U.S. Department of Commerce. https://www.census.gov/library/stories/state-by-state/arkansas-population-change-between-census-decade.html

U.S. Census Bureau, & America Counts Staff. (2021c). *Georgia among top five population gainers last decade*. U.S. Department of Commerce. https://www.census.gov/library/stories/state-by-state/georgia-population-change-between-census-decade.html

U.S. Census Bureau, & America Counts Staff. (2021d). *Louisiana's population was 4,657,757 in 2020*. U.S. Department of Commerce. https://www.census.gov/library/stories/state-by-state/louisiana-population-change-between-census-decade.html

U.S. Census Bureau, & America Counts Staff. (2021e). *Mississippi's population declined 0.2%*. U.S. Department of Commerce. https://www.census.gov/library/stories/state-by-state/mississippi-population-change-between-census-decade.html

U.S. Census Bureau, & America Counts Staff. (2021f). *Tennessee population near 7 million in 2020*. U.S. Department of Commerce. https://www.census.gov/library/stories/state-by-state/tennessee-population-change-between-census-decade.html

U.S. Department of Health and Human Services, HHS. (2020). Defining rural population. Health and Human Services. https://www.hhs.gov/guidance/document/defining-rural-population

U.S. Department of Health and Human Services, HHS, Office of Disease Prevention and Health Promotion. (n.d.). *Social determinants of health*. OASH. https://odphp.health.gov/healthypeople/priority-areas/social-determinants-health

Weldon Cooper Center for Public Service. (2024). *National and 50-state population projections*. University of Virginia.

White, E. M., Morzillo, A. T., & Alig, R. J. (2009). Past and projected rural land conversion in the US at state, regional, and national levels. *Landscape and Urban Planning, 89*(1–2), 37–48.

World Health Organization, WHO. (1948). *Summary Reports on Proceedings Minutes and Final Acts of the International Health Conference held in New York from 19 June to 22 July 1946*. https://iris.who.int/handle/10665/85573

World Health Organization, WHO. (n.d.). *Social determinants of health*. https://www.who.int/health-topics/social-determinants-of-health#tab=tab_1

Chapter 2
History of Rural Health

Abstract The westward expansion of settlers gradually reshaped the landscape of the Southeastern United States. Rural regions, once scattered with sparse buildings, evolved into bustling towns and eventually into thriving cities. This progression facilitated the admission of several states into the Union (i.e., a period of territorial expansion): Tennessee in 1796, Louisiana in 1812, Mississippi in 1817, Alabama in 1819, and Arkansas in 1836. Urban centers emerged rapidly during this time, leading to the establishment of state government buildings, land grant universities, and medical colleges. While medical training was becoming formalized and professional licensing was gradually instituted, early medical practices were largely observational with physicians often undertaking a variety of roles. In rural areas, healthcare was limited and depended heavily on folk remedies or home-based treatments. This chapter offers an interesting historical account of the early practices of rural health.

Keywords Southeastern United States · Westward expansion · Health · Rural regions · Urban centers · Rural health

Introduction

An examination of territorial expansion (i.e., growth of the United States) can enhance our understanding of the status of rural health in the Southeast. The westward migration required settlers to traverse vast areas of land that were often secluded and distant from established towns, leading to a lack of access to medical facilities. Consequently, there emerged a pressing demand for healthcare services to be readily available to these newly inhabited rural regions. The movement of settlers and pioneers resulted in the propagation of contagious diseases, necessitating an understanding of rural health in order to implement preventive measures against ailments such as cholera, smallpox, and typhoid that spread rapidly among these

A. Sharma, *Current and Emerging Rural Health Issues in the Southeast United States*, Global Perspectives on Health Geography,
https://doi.org/10.1007/978-3-032-07543-7_2

expanding communities (Fogel et al., 1978; Higgs, 1973). The westward expansion further illuminated the striking disparity in health outcomes between rural and urban areas. This period highlighted the role of social determinants of health, such as the availability of clean water, sanitation, education, and other socio-economic conditions in shaping the health outcomes of rural populations. Lastly, the westward expansion engendered the development of transportation and infrastructure systems such as roads, railroads, and telegraph lines in order to establish connections between these remote regions and the rest of the nation (Porter, 2024). These advancements facilitated the transportation of medical professionals, supplies, and the delivery of healthcare services to rural areas. Overall, the westward expansion exposed the distinct healthcare challenges faced by rural communities, prompting a deeper understanding of rural health issues and the need for targeted interventions in healthcare.

Historical Overview of Westward Expansion

After the arrival of European settlers in the greater Mid-Atlantic region during the eighteenth century, a noticeable process of expansion began as villages and small settlements underwent the transition into towns (Paxson, 2024). This was primarily driven by population growth and expanded economic opportunities through trade and commerce with Indigenous groups, encompassing activities such as crop cultivation, fishing, and fur trading. These resources played a role in facilitating the development of towns. Industries such as agriculture and textile manufacturing emerged, thereby providing employment options for the local population (Smith, 2024). Meanwhile, the burgeoning commerce and trade fostered the establishment of markets, shops, and various small businesses.

As these towns progressively expanded, the development of essential infrastructure became necessary. This encompassed the construction of transportation networks, including roads, bridges, and later railroads, which not only facilitated trade and the exchange of goods but also enhanced communication among different regions. Access to these transportation networks further promoted the transformation process, solidifying the shift from village to town. Simultaneously, the creation of public amenities such as churches, mills, schools, and other communal facilities became necessary to address the economic and social needs of the growing population and to ensure the provision of adequate support for their needs (Schweitzer, 2000).

During the course of the transition, the establishment of formal local government structures became evident. In the initial stages, villages frequently operated under informal leadership or governance mechanisms. However, as population density increased and administrative demands became more complex, towns progressively institutionalized governance through the election of officials, the enactment of local ordinances, the creation of judicial courts, and the provision of essential public services, including law enforcement and education. These developments facilitated more effective local administration, thereby enhancing self-governance and contributing to the emergence of a more structured and organized community framework.

The westward expansion of early settlers was driven by a variety of factors, including the previously mentioned economic opportunities, the desire for land ownership, and the pursuit of new territories (Wells, 2000). The appeal of fertile land, combined with the potential for increased agricultural output and commercial advancement, shaped settlement patterns during this period. The United States government negotiated treaties and made acquisitions with foreign nations and Native American tribes, often resulting in the relocation of Indigenous populations and the opening of vast territories for settlement. Examples include the Louisiana Purchase (1803), the Indian Removal Act (1830), the subsequent Trail of Tears (1830 to 1850), and the annexation of Texas (1845). As settlers established communities and the population grew in various territories, they sought statehood to gain representation and access to resources. The process of achieving statehood involved demonstrating a certain population size, developing a stable government, and meeting other criteria set by Congress (Onuf, 2000). Figure 2.1 provides a visual depiction of land acquisitions from the late 1700s to mid-1800s. For a century, the United States actively pursued territorial expansion by purchasing land from France, Mexico, and Russia. While Georgia (1788) was already part of the statehood, the westward expansion added Tennessee in 1796, Louisiana in 1812, Mississippi in 1817, Alabama in 1819, and Arkansas in 1836. A dynamic visual can be seen at https://www.visualcapitalist.com/us-territorial-expansion/.

Fig. 2.1 United States expansion. (Source: U.S. Department of Interior, U.S. Geological Survey)

Due to the previously mentioned factors, the Mid-Atlantic region observed considerable population growth (Schweitzer, 2000). Philadelphia, PA had a population of 41,000 while Boston, MA had 25,000 residents, and Charleston, SC had 19,000 residents in 1800 (Carr, 2000; Collins, 2023; Morgan, 2011). New York City surpassed all other cities, reaching a population of over 60,000. By 1850, New York City's population increased rapidly to a staggering 515,000 residents while Boston had almost 137,000 residents and Philadelphia had approximately 121,000 residents. These towns emerged as major urban centers in the mid-1800s and exhibited continuous population growth (Wells, 2000). Similar growth patterns were observed in other parts of the United States. For instance, New Orleans, LA gained prominence in 1850 as the number of residents increased to 117,000, a considerable expansion from nearly 20,000 residents in 1810. Other cities in the Southeast region, such as Memphis, TN, Mobile, AL, and Atlanta and Savannah, GA, also experienced sizable population growth. For a more detailed examination of population growth in the study area from 1850 to 2030, readers are encouraged to refer to Chap. 1.

Readers may also be interested to learn the South was not only an area with a growing population but also functioned as an economic hub of agriculture and trade (Perkins, 2000). For example, cotton served as a valuable cash crop (i.e., crops grown for commercial use) and remains so to this day (see Chap. 1). With the invention of the cotton gin in the late eighteenth century, cotton production increased, leading to a boom in the industry. This resulted in increased profits for planters and further expansion of cotton plantations. The institution of enslavement and indentured servitude played a role in the economic vitality of the South. Laborers provided the workforce for the expansive plantations and elaborate estates that characterized the area, particularly in the cultivation of cotton and tobacco (Rozbicki, 2000). The profitability of the region's agricultural economy heavily depended on labor.

Geography also played a role in the economic dynamics of the South, not only in terms of weather but also its strategic location. The South had several major ports along its coastline, including Charleston, SC, New Orleans, LA, and Savannah, GA. As highlighted in earlier sections of this chapter, these early settlements transformed into major cities due to population growth and economic factors. The Port of Charleston, which connects to the Atlantic Ocean, and the Port of New Orleans, which leads to the Gulf Coast, served as major hubs for facilitating trade and commerce with both domestic and international markets. Goods produced in the region, such as cotton, rice, and tobacco, were exported to Europe and other parts of the United States, contributing to the region's economic growth. Additionally, the South invested in developing transportation infrastructure during the 1800s. This was in the form of canals and railroads which improved connectivity within the region, thereby making it easier to transport goods to markets. The economic prosperity of the South remained tied to agriculture. However, this did not translate to a higher quality of life for those involved in the agrarian economy. Economic disparities were evident between rich landowners and poor indentured workers and enslaved people.

Early Developments in Rural Health

Inequities in health were also evident in many rural parts of the Southeast during the nineteenth century. Urban areas, such as New Orleans, LA, Nashville, TN, Savannah, and Augusta, GA, had better access to care due to the development of hospitals and medical colleges. Established in 1736, Charity Hospital in New Orleans (originally named Hospital of Saint John) is considered among the oldest continuously operated hospitals in the United States. It served both free and enslaved individuals and provided medical care during epidemics such as yellow fever. Candler Hospital, founded in 1804 in Savannah, is the second oldest hospital in continuous operation and was also established to provide housing for those with minimal economic resources. Founded in 1828, the Medical College of Georgia in Augusta is one of the oldest medical schools in the Southeast (see Fig. 2.2). It has a long history of education and research. Tulane University School of Medicine, established in 1834 in New Orleans, is also an early medical college with significant contributions in the areas of yellow fever and cholera.

While hospitals and medical colleges were commonly located in urban areas, their presence did not ensure the delivery of quality healthcare services. The field of medicine was not standardized or regulated (Grabo, 2000). Medical education relied heavily on subjective experiences as opposed to scientific evidence and physicians obtained training through apprenticeships (Hill & Mathews, 2021). Treatments were often determined by tradition and even superstition. Physicians also assumed

Fig. 2.2 Old Medical College in Augusta, GA. (This picture shows the Medical College of Georgia, which was founded in 1828. Source: Historic American Buildings Survey, C. & Clusky, C. (1933). Library of Congress, call number HABS GA,123-AUG,11- [P&P], digital identification: https://hdl.loc.gov/loc.pnp/anrc.03900 Historic American Buildings Survey, C. & Clusky, C. (1933))

multiple roles, including providing general medical care and attending to obstetric needs. In addition to an inadequate understanding of diseases, healthcare was characterized by segregation and inequality. African Americans, both enslaved and free, faced barriers in accessing medical care. Segregated medical facilities frequently provided substandard care to African American patients. Quality care was also inadequate for those with limited economic resources (Ziller & Milkowski, 2020). In addition to access, infectious diseases were rampant during this period, leading to high mortality rates. Epidemics such as cholera, malaria, and yellow fever were particularly devastating (Sullivan, 2010). These diseases proliferated due to inadequate sanitation practices and dense living conditions (Fissell, 2000).

Despite these challenging circumstances, healthcare access and health outcomes were notably worse in rural areas compared to urban areas (Ziller & Milkowski, 2020). Urban centers offered better opportunities for medical education and training as medical schools and universities were primarily located in cities. This enabled urban residents to pursue medical careers. Additionally, urban areas attracted more medical professionals, creating a thriving community with continuous learning opportunities (Grabo, 2000). In contrast, rural areas lacked these educational and training resources, leading to a shortage of healthcare providers. Furthermore, the healthcare infrastructure in rural areas was underdeveloped compared to urban areas. Urban centers had better-equipped medical facilities, greater access to medical technologies, and a larger pool of doctors. Conversely, rural areas had fewer healthcare facilities, limited medical equipment, and a shortage of qualified medical professionals (Ziller & Milkowski, 2020). Lastly, there was often a notable disparity in economic development between rural and urban areas. This economic disparity translated into limited investment in healthcare infrastructure, leading to a scarcity of healthcare facilities and the provision of lower-quality medical services (Ziller & Milkowski, 2020). Consequently, rural inhabitants had to cover extensive distances in order to access medical care.

Delivering Healthcare in Rural Areas

Many rural areas were underserved with a limited number of physicians available to provide care (Sullivan, 2010). In response to these issues, communities often relied on local or folk healers for their healthcare needs. These practitioners delivered basic healthcare services to rural populations. They would treat common ailments and offer guidance on health and hygiene practices. Among the various treatments employed by these local healers was bloodletting, a practice that has garnered considerable historical debate concerning its efficacy and safety (Fissell, 2000). Inspired by early medical practices, this procedure involved intentionally bleeding a person, either by leeches or by making small incisions, in an attempt to balance the body and improve health. Some physicians also adopted this method as a therapeutic option for acute pericarditis and pneumonia (Bryan, 1964). In addition to local healers, spiritual guides offered faith healing along with remedies. In many rural areas of the Southeastern United States, individuals relied on faith healers as an alternative form of treatment (Tannenbaum, 2012). These faith healers offered prayers, placed hands on the body, anointed oils, or performed spiritual rituals to facilitate healing.

Midwives, or traditional birth attendants, played a vital role in rural communities during the nineteenth century. They were responsible for assisting women in childbirth and providing postnatal care. Midwives relied on their experience and knowledge of traditional practices and techniques (Fissell, 2000). They often utilized warm compresses and massages to ease labor pain and promote safe deliveries (Tannenbaum, 2021). In some rural areas, individuals known as “circuit riders” or traveling doctors would periodically visit communities to provide medical care (see Fig. 2.3). These circuit riders would travel by horseback or carriage, treating various ailments and performing basic surgeries (January, 2009). However, their visits were infrequent and not always accessible to everyone in rural areas. In cases where medical attention was urgently required, neighbors and community members worked together to provide assistance. They would often rely on their own limited medical knowledge (Ziller & Milkowski, 2020).

Despite the limitations, efforts were made to improve healthcare services in rural areas during the nineteenth century. Medical colleges and universities started training more doctors and healthcare providers to address the shortages. Additionally, benevolent societies and charities worked towards establishing local clinics and hospitals near rural communities to bring medical services closer to residents. One noteworthy example was the Ladies Benevolent Society of Charleston, SC. This was one of the earliest examples of organized home care. This society primarily served White women. However, it also extended assistance to individuals with limited financial means and independent African Americans. Notably, this initiative spanned from 1813 until the outbreak of the Civil War in 1861, representing a significant period of time where basic healthcare services were accessible to a varied population.

Early Rural Health Practices

In instances where circuit riders or nurses were not accessible to provide essential healthcare services, rural communities relied on home remedies as a primary means of addressing their healthcare needs. These remedies involved utilizing readily available natural ingredients such as herbs, plants, and common household items to address various ailments (Clarke, 1876). For example, feverfew was frequently employed to relieve fever and headaches while elderberry was believed to enhance the immune system and alleviate symptoms of colds and the flu. Poultices made from herbs, vegetables, or clay were applied to wounds, bruises, or painful joints for relief (Meade, 1965). Onion poultices were thought to possess antiseptic and anti-inflammatory properties. To create this poultice, onions were chopped, heated, and wrapped in a cloth before being applied to swollen joints or skin infections to reduce

Fig. 2.3 Rural nurse. This picture shows a rural nurse navigating the landscape by horseback. Her attire, which comprises a Western-style hat, a heavy winter coat, and riding boots, reflects an adaptation to the environment of rural settings. Additionally, a medical bag is slung over her shoulder, indicating her role as a healthcare professional. Source: American Red Cross, 22 August 1919. Library of Congress, call number LC-A6195–6524 [P&P], digital identification: https://hdl.loc.gov/loc.pnp/anrc.03900

pain and inflammation (Cavender, 2014). Compresses and bandages made from cloth were also utilized to cover and protect wounds. Mustard plasters, for example, were made by mixing powdered mustard with water and spreading the paste on cloth (Warner, 2014). The cloth would then be applied to the chest to alleviate congestion caused by respiratory illnesses such as colds or bronchitis.

Another category of home remedies constituted medicinal teas and tonics. Infusions and decoctions made from plants and herbs were frequently employed as remedies (Tannenbaum, 2012). Chamomile tea, known for its calming properties, was utilized to ease digestion and promote relaxation. Sassafras root bark was often boiled to produce a tea allegedly possessing blood-purifying properties. This tea was believed to aid in the treatment of skin conditions such as acne or eczema while also alleviating rheumatic pain. Catnip, recognized for its calming effects on cats, was also utilized in humans as a mild sedative. It was commonly brewed into an infusion or tea for consumption, especially to aid with sleeplessness or indigestion.

Folk remedies were employed to address everyday health concerns. Honey and lemon, for instance, were commonly used to soothe sore throats while a hot drink made with whiskey or bourbon, honey, lemon, and hot water was believed to alleviate cold symptoms and congestion. Tobacco smoke concoctions were employed in the treatment of various medical conditions, particularly respiratory ailments such as asthma and croup. The underlying belief was that nicotine would stimulate respiratory function (Haynes, 2012; Lawrence, 2002). In some cases, this treatment was also used to provide relief from colds.

Advancements in Rural Healthcare

The provision of rural healthcare in the nineteenth century Southeastern United States was particularly challenging for rural communities. Consequently, disparities in healthcare access and quality between urban and rural areas were pronounced (Ziller & Milkowski, 2020). Nevertheless, these disparities were not limited to that period as they have persisted over the years. Efforts have been undertaken to bridge this gap and improve healthcare access for rural communities with one notable example being the New Deal initiatives.

During the Great Depression in the 1930s, the nation faced significant economic hardships, leading to a lack of funding and resources, particularly in rural areas. This resulted in low levels of public confidence and sentiment. However, President Franklin D. Roosevelt (office from 1933–1945) and his New Deal programs brought about transformative changes in the country. One such initiative was the Rural Electrification Administration (REA), which was authorized by Congress in 1936 (Carmody, 1939). This program allocated funds to local cooperatives for the purpose of distributing electricity to rural regions.

Prior to the 1940s, almost 90% of farms in rural areas lacked access to electricity. This deficiency meant that water had to be manually drawn from pumps, there were no electric lights, radios, or telephones, and refrigeration for food was non-existent (Lewis, 2018). The introduction of electricity facilitated the establishment of updated facilities, leading to improvements in indoor plumbing and access to household appliances. These developments contributed to enhanced sanitation and improved overall health conditions in many rural farms (Karr-Kaitin, 2013). Additionally, the availability of electricity in rural areas facilitated a safer working environment for farmers by reducing their dependence on kerosene lamps, thereby minimizing exposure to hazardous particulate matter (Rovang, 2016). Furthermore, electricity granted farms access to communication channels and entertainment options. As emphasized in the preceding chapter, health encompasses mental and social well-being (see WHO definition in Chap. 1). The advent of electrification enabled rural households to connect through telephones and radios, alleviating their isolation and loneliness. Telephones allowed instant communication with friends, family members, and businesses in distant towns and cities. Additionally, telephones provided a means for rural communities to connect with medical professionals, enabling them to seek medical advice without having to travel long distances (Mermelstein et al., 2017). Radio played a central role in connecting rural households to the outside world, as well. With access to radio broadcasts, rural communities could obtain news, information, and music. The radio became a source of cultural enrichment, education, and entertainment for rural households (Craig, 2001). It allowed residents to stay updated on the latest events and participate in community events such as radio shows and contests (Wik, 1981). Consequently, the social fabric of rural life experienced an enhancement through increased connectivity. By 1950, nearly 40% of the nation's agricultural establishments had been provided with electricity.

In the post-World War II era, the Southeastern United States witnessed considerable improvements in rural health infrastructure. The Hospital Reconstruction Act of 1946, which will be explored comprehensively in Chap. 3, provided funding for the construction of hospitals and other healthcare facilities. Recognizing that providing accessible and quality healthcare services was essential for the overall well-being and development of the nation, this legislation aimed to address the existing inadequacies in rural health facilities through financial investments. By allocating federal funds towards rebuilding and establishing hospitals and healthcare centers, the act bridged the healthcare gap between rural and urban areas while ensuring equitable access to healthcare services. Additionally, the infusion of financial resources allowed for the modernization of healthcare infrastructure, enabling the adoption of advanced medical equipment and enhanced diagnostic capabilities. As a result, rural communities not only experienced an increase in the quantity of healthcare facilities but also an improvement in the quality of services offered.

In recent decades, community health centers and rural health clinics have emerged as central components of addressing healthcare disparities in rural areas. These healthcare facilities were established with the goal of improving the accessibility of services for underserved rural populations. Specifically, rural health clinics serve as primary care centers that offer healthcare services and are strategically located in medically underserved areas. Later in this chapter, I will expound on health professional shortage areas and underserved communities.

Rural health clinics employ a diverse team of healthcare professionals, such as nurse practitioners, physicians, physician assistants, and support staff (U.S. HHS, 2007). These clinics provide a wide range of services that encompass critical care for minor illnesses and injuries, chronic disease management, preventive care, and basic laboratory services but do not have a minimum service requirement. In some instances, these facilities establish collaborations with local hospitals and healthcare systems to ensure a continuum of care, thereby ensuring that patients can access specialized services when required. In contrast, community health centers constitute more comprehensive healthcare facilities that serve both rural and urban communities. Although they are not exclusively designed for rural areas, these centers have played a role in expanding healthcare access in underserved communities, including rural regions in the Southeast. Community health centers offer diverse services, encompassing preventive care, primary care, dental care, and mental health services. Moreover, they may adopt a team-based approach to care with physicians, nurse practitioners, nurses, dentists, mental health professionals, and other healthcare providers by collaborating to meet the multifaceted needs of patients. These centers may also offer health education and outreach programs. Such programs focus on educating rural populations about disease prevention, hygiene, and proper nutrition, leading to improved overall health outcomes.

Figure 2.4 illustrates the distribution of rural and urban counties within the study area. This map employs a key categorical representation, wherein rural regions are depicted in green and urban regions in yellow. The intensity of the shading serves as an indicator of the degree of either rurality or urbanity (i.e., predominant category). Specifically, darker shades of green signify greater levels of rurality while darker shades of yellow reflect higher levels of urbanization. Conversely, lighter shades of green or yellow denote lower levels of rural or urban characteristics, respectively. Placing this map adjacent to Fig. 2.5, depicting the Health Professional Shortage Area scores, facilitates a clearer visualization of the shortage of primary care workers in rural regions. The HPSA primary care score assigns a numerical value to prioritize the allocation of healthcare professionals in primary care areas (HHS, 2022). Ranging from 0 to 25, higher values correspond to a greater need and are perceptibly shaded in darker blue. Upon examination of both maps, readers may readily observe a significant spatial overlap between extensive dark green areas in Fig. 2.4 and expansive dark blue areas in Fig. 2.5 (i.e., areas southeast of Little Rock, AR, outside Baton Rouge and New Orleans, LA, south of Birmingham, AL, west of Nashville, TN, and south of Atlanta, GA). Rural communities typically

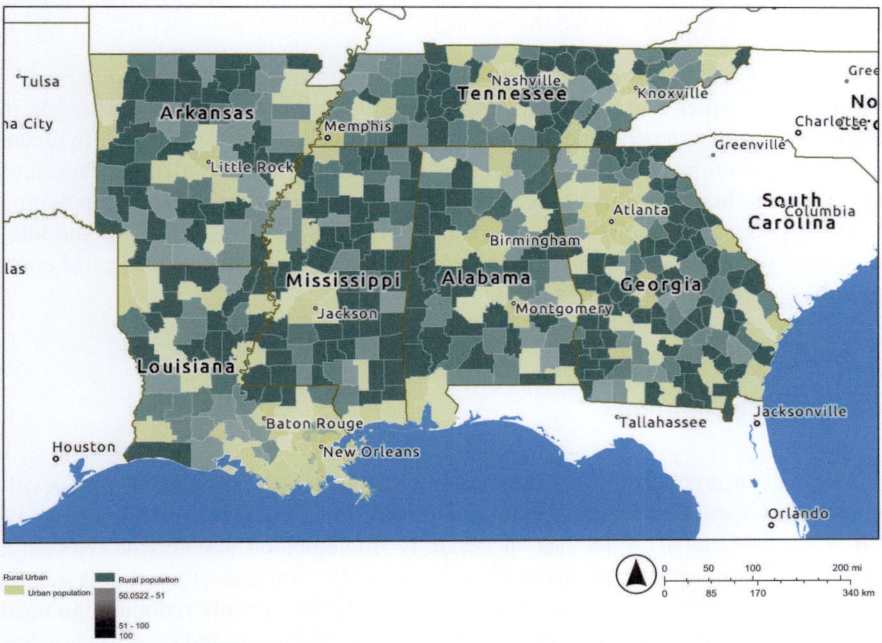

Fig. 2.4 Rural and urban distribution. (Source: ESRI, TomTom, Garmin, FAO, NOAA, USGS, EPA, USFWS (A. Sharma))

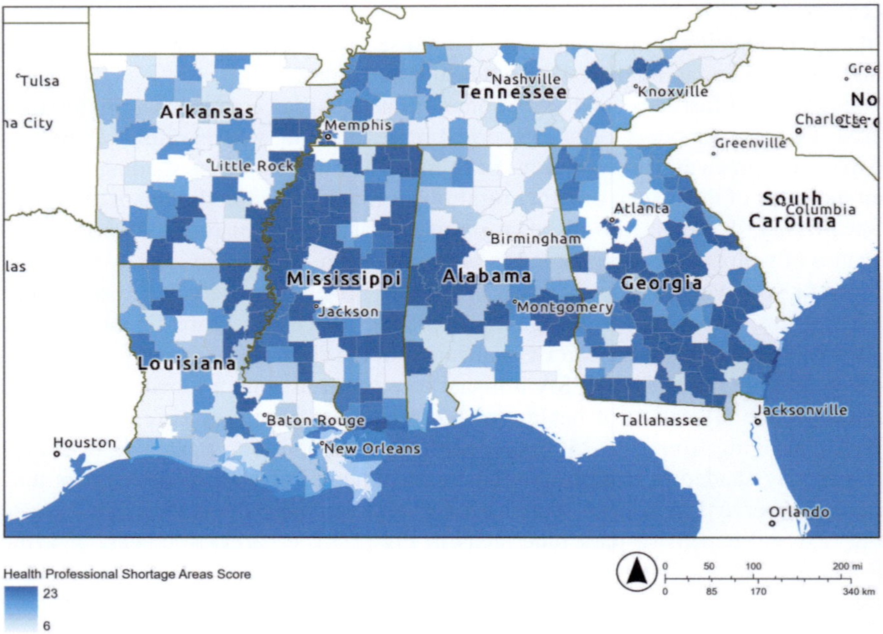

Fig. 2.5 Health Professional Shortage Area (HPSA) score. (Source: ESRI, TomTom, Garmin, FAO, NOAA, USGS, EPA, USFWS (A. Sharma))

score at the higher range for the HPSA primary care score due to challenges in attracting healthcare personnel such as nurses, physicians, and specialists. In recent years, several initiatives have been undertaken to address these challenges. Programs to retain healthcare professionals in rural areas have been implemented, including loan repayment programs for rural practitioners. Additionally, telehealth and telemedicine services have gained prominence, bridging the gap between rural communities and healthcare providers.

Recent Developments

Current advancements in rural healthcare have centered around technological developments in computing and telecommunications. For instance, the University of Mississippi Medical Center has successfully implemented a statewide telehealth network. This innovative program allows patients in rural areas to remotely access specialty care, including telepsychiatry services, thereby greatly reducing the need for travel (deShazo & Parker, 2017). Similarly, the Alabama Medicaid Agency has introduced the Alabama Coordinated Health Network (ACHN) initiative, which is aimed at improving health outcomes for Medicaid recipients and increasing access to care in rural regions by utilizing telemedicine, mobile health units, and community health workers. ACHN aims to deliver preventive care, manage chronic diseases, and offer behavioral health services to underserved populations with guidance and collaboration from the Alabama Department of Public Health (Alabama Medicaid Agency, 2023).

Telehealth is a comprehensive concept that encompasses the use of digital technologies to deliver healthcare services remotely (HRSA, 2022). These services may include virtual doctor visits, remote patient monitoring, and electronic transmission of medical records. Telehealth leverages mobile communication applications, remote monitoring devices, and video conferencing to connect patients with healthcare professionals. In contrast, telemedicine specifically refers to the practice of providing medical care remotely through telecommunication technologies (FCC, 2023). It encompasses the diagnosis, treatment, and monitoring of patients, featuring virtual consultations, remote vital sign monitoring, and secure electronic platforms for the sharing of medical information.

The following example pertains to recent legislative endeavors proposed by Congress, specifically focusing on the Preserving Telehealth, Hospital, and Ambulance Access Act or PTHAAA. Introduced by Representatives David Schweikert (AZ) and Mike Thompson (CA) as H.R. 8261 in May 2024, the bill aims to preserve and expand access to telehealth services (PTHAAA, 2024). One key provision is the amendment of Title XVIII of the Social Security Act to expand Medicare coverage for telehealth services. This expansion would allow Medicare beneficiaries to access a wider range of healthcare services through telehealth, including mental health counseling, preventive care, and specialist consultations. Furthermore, the bill proposes to remove geographic and site restrictions that were in place before the coronavirus disease of 2019, thereby enabling patients to receive telehealth services from their homes. Although the bill did not receive approval during the December 2024 legislative session, members of Congress may modify its provisions and reintroduce it in a future session. Given the ongoing emphasis on reducing healthcare expenditures by both state and federal governments, an alternative version of this bill will likely be proposed.

The success of any future legislative proposal rests on the implementation of the Internet for All initiative, which plays an important role in bridging the digital divide. This program provides funding for technological equipment and broadband internet access, particularly in underserved areas lacking the necessary resources to access business, educational, and telehealth services (NTIA, 2022). By expanding telehealth infrastructure, the Internet for All initiative empowers individuals residing in rural areas to conveniently access healthcare services without facing the traditional barriers they may encounter. Notably, several states, including Alabama, Arkansas, Georgia, Louisiana, and Tennessee, have actively participated in the Broadband Equity, Access, and Deployment (BEAD) program. Through their involvement, these states have demonstrated a commitment to enhancing broadband access and ensuring equitable distribution of resources. For more information regarding this program, visit the official website at https://www.internetforall.gov or https://broadbandusa.ntia.doc.gov.

Concluding Remarks

Overall, the history of rural health in the Southeast United States highlights a gradual improvement in healthcare infrastructure and services. However, ongoing efforts and investments are necessary to ensure equitable access to quality healthcare for rural communities in the region (Kolluri et al., 2022). Rural areas in the Southeast often lack adequate healthcare resources, including hospitals, clinics, and specialists. Geographical isolation and a sparse population make it difficult for healthcare providers to establish and sustain practices, leading to diminished access to essential healthcare services (Andrews et al., 2020). One of the primary barriers to rural health in the Southeast involves shortages of medical professionals. Rural areas in the Southeast also struggle with poverty and limited resources, affecting individuals' ability to afford or access healthcare and other services (Peluso et al., 2024). Unemployment rates, higher rates of uninsured individuals, and low-income levels contribute to barriers in obtaining essential medical services (Streeter et al., 2020). The previously mentioned technological advancements in healthcare demonstrate the potential to address the disparity in healthcare access and delivery. However, rural areas in the Southeast often have limited access to high-speed internet and technological infrastructure, delaying the adoption of innovative healthcare practices (Sharma, 2023). While the previously mentioned Internet for All initiative is a noteworthy effort to improve digital access to telehealth programs, developing advanced programs may not be sufficient. A recent study found that Black patients were less likely to use telehealth compared to White patients, potentially exacerbating existing healthcare disparities (Bakaki et al., 2021). Similarly, research has shown that machine learning algorithms used for analyzing medical images can be biased based on race and other demographics. These algorithms display less accuracy in identifying health conditions in patients from underrepresented racial and ethnic groups, potentially leading to disparities in diagnosis and treatment (Kornfield et al., 2020). In an upcoming chapter, I will provide a succinct overview of the initiatives undertaken by the American Medical Association in response to these concerns. Additionally, I will discuss the significance of exercising voting rights within a democratic framework that acknowledges and addresses civic participation as one approach to address health inequities. In conclusion, healthcare practitioners, policymakers, business leaders, and community workers must continue to collaborate and propose creative solutions to reduce the barriers to health equity (Soltani et al., 2023). I trust that this book will encourage dialogue and inspire efforts to overcome these challenges.

References

Alabama Medicaid Agency. (2023). *Alabama Coordinated Health Network (ACHN) Quality Strategy*. Alabama Medicaid Agency. https://medicaid.alabama.gov/documents/5.0_Managed_Care/5.1_ACHN/5.1.4_ACHN_Administration/5.1.4_Proposed_Draft_ACHN_Quality_Strategy_7-11-23.pdf#:~:text=services%20are%20still

Andrews, M. R., Tamura, K., Claudel, S. E., Xu, S., Ceasar, J. N., Collins, B. S., Langerman, S., Mitchell, V., Baumer, Y., & Powell-Wiley, T. M. (2020). Geospatial analysis of neighborhood deprivation index (NDI) for the United States by county. *Journal of Maps, 16*(1), 101–112.

Bakaki, P. M., et al. (2021). Association of race and ethnicity with health care use and outcomes in COVID-19. *JAMA Network Open, 4*(3), e211519.

Bryan, L. S. (1964). Blood–Letting in American Medicine, 1830–1892. *Bulletin of the History of Medicine, 38*(6), 516–529.

Carmody, J. M. (1939). Rural electrification in the United States. *The Annals of the American Academy of Political and Social Science, 201*(1), 82–88.

Carr, J. B. (2000). A change "as remarkable as the revolution itself": Boston's demographics, 1780–1800. *The New England Quarterly, 73*(4), 583–602.

Cavender, A. (2014). *Folk medicine in Southern Appalachia*. UNC Press Books.

Clarke, E. H. (1876). *A century of American medicine, 1776–1876*. Lea.

Collins, S. (2023). A visualization tool for 1790s Charleston: Locating an enslaved population using GIS. *Journal of Urban History, 49*(6), 1342–1357.

Craig, S. (2001). " The farmer's friend": Radio comes to rural America, 1920–1927. *Journal of Radio Studies, 8*(2), 330–346.

deShazo, R. D., & Parker, S. B. (2017). Lessons learned from Mississippi's telehealth approach to health disparities. *The American Journal of Medicine, 130*(4), 403–408.

Federal Communications Commission, FCC. (2023). *Telehealth, telemedicine and telecare: What's what*. Federal Communications Commission. https://www.fcc.gov/general/telehealth-telemedicine-and-telecare-whats-what

Fissell, M. E. (2000). Medicine before and after the revolution. In *A companion to the American revolution* (pp. 600–604). Wiley.

Fogel, R. W., Engerman, S. L., Trussell, J., Floud, R., Pope, C. L., & Wimmer, L. T. (1978). The economics of mortality in North America, 1650–1910: A description of a research project. *Historical Methods: A Journal of Quantitative and Interdisciplinary History, 11*(2), 75–108.

Grabo, N. S. (2000). The cultural effects of the revolution. In *A companion to the American revolution* (pp. 586–594). Wiley.

Haynes, S. (2012). Tobacco smoke enemas. *British Columbia Medical Journal, 54*(10), 496.

Health Resources and Service Administration, HRSA. (2022). *What is telehealth?* HRSA. https://www.hrsa.gov/telehealth/what-is-telehealth

Higgs, R. (1973). Mortality in rural America, 1870-1920: Estimates and conjectures. *Explorations in Economic History, 10*(2), 177.

Hill, C. E., & Mathews, H. (2021). Traditional health beliefs and practices among southern rural blacks: A complement to biomedicine. In *Perspectives on the American south* (pp. 307–322). Routledge.

Historic American Buildings Survey, C, & Clusky, C. (1933). *Old Medical College, 598 Telfair Street, Augusta, Richmond County, GA*. Richmond County Augusta Georgia. Documentation Compiled After 1933. [Photograph] Retrieved from the Library of Congress from https://www.loc.gov/item/ga0178/

January, A. M. (2009). Friday at frontier nursing service. *Public Health Nursing, 26*(2), 202–203.

Karr-Kaitin, N. (2013). *How it worked when it worked*. Electrifying Rural America.

Kolluri, S., Stead, T. S., Mangal, R. K., Coffee, R. L., Jr., Littell, J., & Ganti, L. (2022). Telehealth in response to the rural health disparity. *Health Psychology Research, 10*(3), 37445.

Kornfield, R., et al. (2020). Bias in medical image analysis: Association with race and ethnicity. *National Academy of Sciences of the United States of America, 117*(4), 1608–1610.
Lawrence, G. (2002). Tobacco smoke enemas. *The Lancet, 359*(9315), 1442.
Lewis, J. (2018). Infant health, women's fertility, and rural electrification in the United States, 1930–1960. *The Journal of Economic History, 78*(1), 118–154.
Meade, G. M. (1965). Home remedies in rural America. *Annals of the New York Academy of Sciences, 120*(3), 823–828.
Mermelstein, H., Guzman, E., Rabinowitz, T., Krupinski, E., & Hilty, D. (2017). The application of technology to health: The evolution of telephone to telemedicine and telepsychiatry: A historical review and look at human factors. *Journal of Technology in Behavioral Science, 2*, 5–20.
Morgan, K. (2011). *British American port cities*. Oxford University Press.
National Telecommunications, & Information Administration, NTIA. (2022). *Internet for all*. United States Department of Commerce. https://broadbandusa.ntia.doc.gov/sites/default/files/2022-06/IFA-Overview-IFA-Launch-Final-Updated.pdf
Onuf, P. S. (2000). The west: Territory, states, and confederation. In *A companion to the American revolution* (pp. 374–382). Wiley.
Paxson, F. L. (2024). *History of the American frontier*. Gildan Media LLC aka G&D Media.
Peluso, A., Tuccillo, J., Sparks, K., Kapadia, A., & Hanson, H. A. (2024). Spatial analysis of social capital and community heterogeneity at the United States county level. *Applied Geography, 162*, 103168.
Perkins, E. J. (2000). Socio-economic development of the colonies. In *A companion to the American revolution* (pp. 51–59). Wiley.
Porter, R. P. (2024). *The west, from the census of 1880. A history of the industrial, commercial, social, and political development of the states and territories of the west from 1800 to 1880*. BoD–Books on Demand.
Preserving Telehealth, Hospital, and Ambulance Access Act, PTHAAA. (2024). *H.R. 8261*. House Bill. https://www.congress.gov/bill/118th-congress/house-bill/8261/titles
Rovang, S. (2016). The grid comes home: Wiring and lighting the American farmhouse. *Buildings & Landscapes: Journal of the Vernacular Architecture Forum, 23*(2), 65–88.
Rozbicki, M. J. (2000). The cultural development of the colonies. In *A companion to the American revolution* (pp. 65–81). Wiley.
Schweitzer, M. M. (2000). The economic and demographic consequences of the American revolution. In *A companion to the American revolution* (pp. 559–578). Wiley.
Sharma, A. (2023). Exploratory spatial analysis of food insecurity and diabetes: An application of multiscale geographically weighted regression. *Annals of GIS, 29*(4), 485–498.
Smith, D. M. (2024). The geographical basis of American life: Space and Society in the Development of the United States. In *The United States* (pp. 1–53). Routledge.
Soltani, S., Hinman, J. A., Blanco-Velazquez, I., Banchoff, A. W., Campero, M. I., Nelson, L. M., & King, A. C. (2023). Bringing micro to the macro: How citizen science data enrich geospatial visualizations to advance health equity. *Journal of Maps, 19*(1), 2216217.
Streeter, R. A., Snyder, J. E., Kepley, H., Stahl, A. L., Li, T., & Washko, M. M. (2020). The geographic alignment of primary care health professional shortage areas with markers for social determinants of health. *PLoS One, 15*(4), e0231443.
Sullivan, G. (2010). Plantation medicine and health care in the old South. *Legacy, 10*(1), 3.
Tannenbaum, R. (2012). *Health and wellness in colonial America*. Bloomsbury Publishing USA.
Tannenbaum, R. (2021). Women and medicine in early America. In *Oxford research encyclopedia of American History*. Oxford University Press.
U.S. Department of Health and Human Services, HHS. (2007). *Rural health clinic fact sheet*. Centers for Medicaid and Medicare Services. https://www.cms.gov/files/document/rhcfactsheetpdf
U.S. Department of Health and Human Services, HHS. (2022). *Scoring shortage designations*. Health Resources and Services Administration. https://bhw.hrsa.gov/workforce-shortage-areas/shortage-designation/scoring

Warner, J. H. (2014). *The therapeutic perspective: Medical practice, knowledge, and identity in America, 1820–1885* (Vol. 371). Princeton University Press.

Wells, R. V. (2000). Population and family in early America. In *A Companion to the American revolution* (p. 33, 39). Wiley.

Wik, R. M. (1981). The radio in rural America during the 1920s. *Agricultural History, 55*(4), 339–350.

Ziller, E., & Milkowski, C. (2020). A century later: Rural public health's enduring challenges and opportunities. *American Journal of Public Health, 110*(11), 1678–1686.

Chapter 3
Social and Racial Considerations for Health

Abstract The history of the Southeastern United States is characterized by a complex interplay of Indigenous presence and European colonization. Indigenous populations inhabited the region for centuries prior to the arrival of French, English, and Spanish settlers in the seventeenth century, whose incursions precipitated the displacement of Indigenous groups from their ancestral lands. Throughout the eighteenth century, colonial settlers advanced entrepreneurial initiatives while increasingly relying on the labor of Indigenous peoples, women, children, and African Americans to facilitate the nation's independence and westward expansion. This period was defined by ideological commitments to self-reliance, the principle of "free suffrage," and the elevation of the "common man," which collectively contested the prevailing notion that property ownership was essential for men to contribute to the nation's economic development. Notwithstanding the formal abolition of slavery following the American Civil War (1861–1865), systems analogous to indentured servitude persisted in various parts of the Southeast. Following the Women's Suffrage Movement (1848–1920) and the Civil Rights Movement (1954–1968), minority groups increasingly attained greater socio-economic, political, and legal equality. This chapter explores the intersection of social and racial factors and their implications for health outcomes within the region.

Keywords Indigenous populations · Early health practices · Social issues and health · Racial issues and health · Equity · Health inequities

Historical Overview of the Southeast

The Southeastern United States presents a rich historical tapestry, reflecting its indigenous roots, colonial history, and contemporary developments. A comprehensive understanding of this historical progression is essential for informed policy-making and future healthcare planning in the region. While individuals may have a cursory understanding of the South's history from grade school or high school

A. Sharma, *Current and Emerging Rural Health Issues in the Southeast United States*, Global Perspectives on Health Geography, https://doi.org/10.1007/978-3-032-07543-7_3

educational settings—such as Sarah and Angelina Grimke's advocacy of women's rights during the 1800s, the Civil War's pivotal events in Chattanooga and Chickamauga, TN during the 1860s, or the march from Selma to Montgomery, AL lead by Martin Luther King, Jr. during the Civil Rights Movement of the late 1950s and 1960s—the comprehensive narrative of the South begins thousands of years in the past (Cooper Jr & Terrill, 2009; Flexner & Fitzpatrick, 1996; Mulligan et al., 2004; Raghavan et al., 2015; Saunt, 2017). This narrative offers a deeper and more nuanced understanding of the region's historical and demographic journey.

The Native Americans were early inhabitants of the area. The Tuskaloosa and Coosa, for example, settled in the Southeastern Woodlands, which is present-day Alabama and Mississippi. In response to the threat of Spanish conquistadors, the Indigenous peoples gradually organized and formed the Choctaw, Creek, and Cherokee during the 1600s. The French also explored the Gulf of Mexico (as of January 2025, referred to as the Gulf of America in the United States) during the late seventeenth century and some expeditions formed alliances with Indigenous groups while others interacted in an acrimonious manner (Jaenen, 2001). Figure 3.1 provides a geographic detail of the various territories controlled by France, Great Britain, and Spain during the 1600s and early 1700s. Readers may want to note the considerable number of forts established by France in the Midwest region. These forts are spread throughout the expanse and served as administrative centers, supply points, and military bases for newly acquired territories.

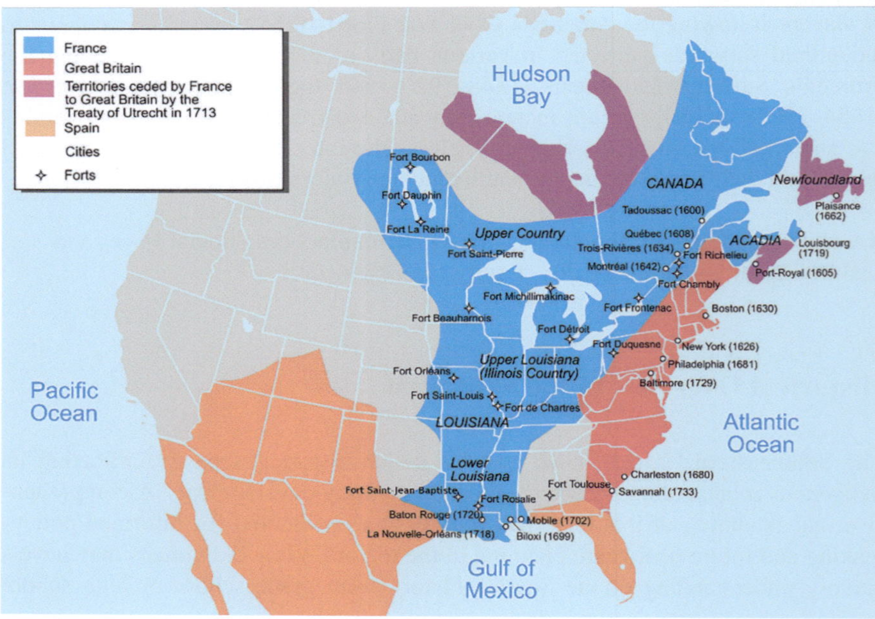

Fig. 3.1 Territories held by France, Great Britain, and Spain (early-1700s). Red symbolizes the English settlements, blue represents the French Territories, and orange depicts areas controlled by Spain. (Source: created on 17 April 2010 and attributed to Pinpin (n.d.))

France occupied most of what is now referred to as the Midwest region, Great Britain held the Northeast and Atlantic region, and Spain occupied the Southwest region and present-day Mexico. France also controlled the northeastern portion of Canada (see forts in Fig. 3.1). Under King Louis XIV, the French Code Noir was promulgated in 1685, leading to the formal institution of enslavement (Palmer, 1995). It was expanded in 1724 into Louisiana and then other French territories. This era saw a burgeoning trade in Africans and a shift from Native American to plantation-based enslavement, driven by the expanding cotton industry. Despite frequent clashes with the English and Indigenous groups during the 1700s, the French retained several territories throughout the region. Fig. 3.2 provides a geographic detail of the territories held two decades before establishment of the American colonies (in yellow) along with English settlements (in red) and French territories (in green).

Despite the United States' victories in the American Revolutionary War (1775–1783) and the War of 1812, the institution of indentured servitude continued to exert an influence on the socio-economic landscape. This system primarily impacted African Americans and impoverished Whites lacking land ownership and it remained a significant element of the Antebellum South (Little, 2022). The continued existence of indentured servitude during this period highlights the enduring structural inequalities that characterized the region, complicating prevailing

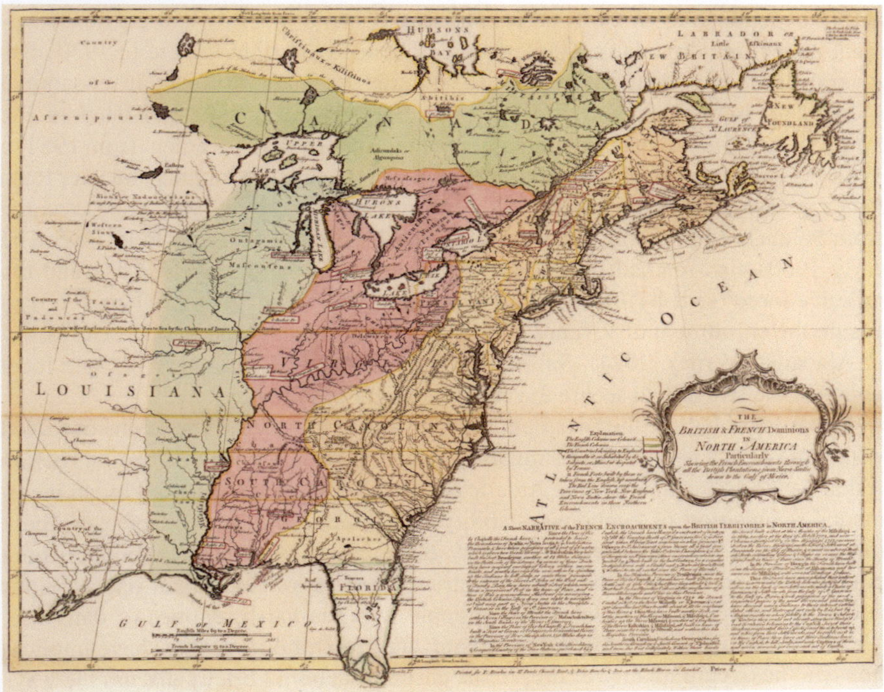

Fig. 3.2 Territories held by France, Great Britain, and Spain (mid-1700s). Yellow denotes the American Colonies, red symbolizes the English settlements, and green represents the French Territories. (Source: created in 1758 and attributed to Bowles, Bowles, & Son)

narratives of democratic advancement and individual liberty that emerged in the aftermath of these seminal conflicts. Illustrations of such democratic progress include the valorization of self-reliance, the principle of "free suffrage," and the political ascendance of the "common man," all of which collectively challenged the well-established assumption that property ownership was necessary for meaningful participation in the nation's economic development.

As discussed in Chap. 2, the westward expansion was also characterized by policies that prioritized settler interests, such as the Indian Removal Act of 1830. The consequent events of the Trail of Tears (1830 to 1850) disrupted Indigenous cultures and facilitated the establishment of a plantation economy that depended heavily on labor. Additionally, the economic reliance on African Americans, women, children, and White men without land ownership further complicated the social hierarchy. These individuals often found themselves in difficult positions, caught between their aspirations for upward mobility and the economic realities that offered few options for improvement. President Abraham Lincoln made an early effort during the American Civil War (1861–1865) with the Emancipation Proclamation, an executive order that changed the status of all enslaved persons to forever free in January 1863. After the Union Army prevailed against the Confederate Army in 1865, the practice of enslavement was abolished and African Americans were granted citizenship with the 13th and 14th Amendments, respectively, and then the right to vote with the 15th Amendment (Rome, 2022). Despite this, African Americans continued to experience restrictions on economic and social mobility and unequal treatment under Jim Crow laws. While women obtained the right to vote in 1920 after several decades of suffrage, they also endured unequal treatment in economic, social, and political arenas. Racial and gender discrimination persisted in the Southeastern United States, culminating in the Civil Rights Movement from 1954 to 1968. Alabama was central to this movement with Claudette Colvin's and later Rosa Parks' refusal to relinquish their bus seats in Montgomery in 1955 and the march from Selma to Montgomery, which was led by Martin Luther King, Jr. in 1965. The early 1960s also marked the conflict between Governor George Wallace and the military force (sent by President John F. Kennedy, office from 1961 to 1963) to ensure the enrollment of Vivian James Malone and James Hood into the University of Alabama (Verney, 2024). Subsequently, Selma and Montgomery became focal points in the struggle to secure the constitutional right to vote.

In the latter half of the twentieth century, a gradual advancement was observed within the African American community, women, and other people of color, such as Asian Americans and Hispanic Americans. This progress was marked by an incremental acquisition of economic stability and political influence. However, these improvements have not been uniformly experienced. A sizable portion of these populations, particularly those with restricted access to educational opportunities and financial resources, continue to grapple with the harsh realities of daily life. These hardships, often a result of systemic and structural inequities, have implications on the overall quality of life. In the ensuing section, I briefly discuss the ways in which these socio-economic factors exert influence on the health and well-being of these populations. The objective is to review the intricate relationship between these elements and how they collectively shape the lived experiences of diverse groups. This understanding is critical in formulating policies and interventions aimed at promoting health equity.

Social Considerations and Health

Social justice is a multifaceted concept that refers to the equitable treatment and status of individuals and groups within society (Miller, 1979). This includes justice in terms of the distribution of opportunities. Specifically, social justice assigns rights and responsibilities within societal institutions and is often associated with the equitable allocation of economic, political, and social advantages and responsibilities. Furthermore, social justice encompasses equitable opportunities for individuals to contribute to and derive advantages from collective welfare. Lastly, social justice highlights the structural conditions that promote individual growth and empower individuals to exercise their autonomy in decision-making. This includes the ability to engage in public service and hold public office. In essence, it is about creating a society where all members have equal access to opportunities and resources and such rights are recognized and protected (Jayasinghe et al., 2022). It is a principle that guides us towards a more equitable and inclusive society.

Indigenous Communities

Social justice remains a concern in the Southeastern United States, particularly affecting Indigenous communities in multiple ways (French, 2003). The historical domination of Indigenous nations by early French and Spanish settlers has resulted in ongoing social disruptions that continue to afflict tribal communities (see Chap. 2). Indigenous people living on reservations encounter unique challenges due to their confinement within the boundaries of allocated lands. As depicted in Fig. 3.3, the Indigenous population resides in specific areas throughout the Southeastern region. In this map, deeper shades represent higher population numbers for the Indigenous communities.

For example, this choropleth map shows Indigenous people (e.g., Quapaw) are concentrated in the western regions of Arkansas while the population is sparse in the northeastern and southeastern regions of the state. In Louisiana, the northeastern part has a smaller population while a sizable portion (e.g., Natchez) resides in the northwestern part of the state and New Orleans. A cluster is also evident in central Mississippi (e.g., Chickasaw and Choctaw). The pattern is more diffuse in Alabama with Indigenous groups in the north, near Birmingham, and the southwestern part of the state (e.g., Cherokee, Choctaw, and Creek). While living on tribal land allows Indigenous peoples sovereignty over the land and natural resources, this geographical limitation can restrict their access to essential resources such as employment

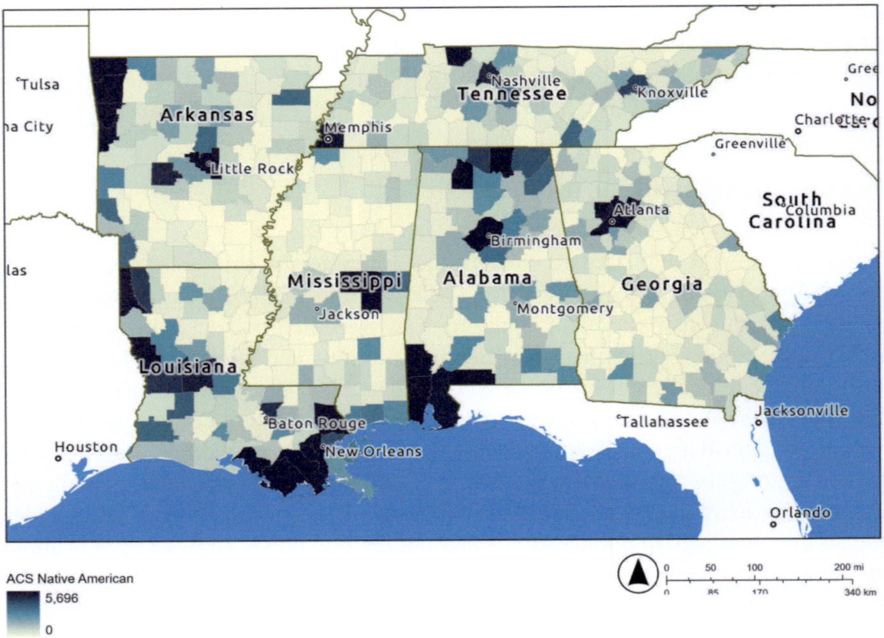

Fig. 3.3 Indigenous population distribution. (Source: ESRI, TomTom, Garmin, FAO, NOAA, USGS, EPA, USFWS (A. Sharma))

opportunities, education, and healthcare. As a result, poverty levels are disproportionately high within these communities. The lack of substantial employment prospects exacerbates the economic struggles faced by Indigenous groups (Akee et al., 2019). Limited job opportunities within reservation boundaries contribute to lower median incomes compared to the national average. Many tribal members can be caught in a cycle of poverty with few avenues for upward mobility. Access to higher education is also limited for Indigenous individuals residing on reservations (Marley, 2023). Geographic isolation, coupled with financial constraints, hinders their ability to pursue college degrees or vocational training. As a consequence, educational attainment remains below the national average.

Food insecurity is another pressing issue among Indigenous populations. Several factors contribute to this. First, the previously mentioned geographical constraints play a role. A comprehensive study revealed that nearly 30% of individuals living on tribal lands are within a mile of a grocery store, in stark contrast to 60% of the overall population (Kaufman et al., 2014; Nikolaus et al., 2022). The scarcity of nearby food sources makes it challenging for Indigenous families to access fresh and nutritious food. Second, economic hardships directly impact food security. The previously mentioned lower incomes and limited employment opportunities make it difficult for these individuals to afford a balanced diet. The consequences of food insecurity extend beyond hunger. Substance abuse, alcoholism, and disproportionately higher rates of health conditions (such as diabetes and obesity) are prevalent within Indigenous communities (Singh et al., 2021). Inadequate access to healthcare services further compounds these health disparities (Kruse et al., 2022; Sommerfeld et al., 2021).

Women

The intersection of social justice and health for women in the Southeast United States is another complex issue that is influenced by several factors such as access to healthcare services, race, and socio-economic status (Andrews et al., 2020; Qiu & Wu, 2022; Sharma, 2022, 2023). The Southeast, which accounts for nearly 10% of the nation's population, is characterized by high rates of chronic disease and poor health outcomes. These health disparities are pronounced among women, who often face barriers to accessing quality healthcare services. One notable example is the maternal health crisis with the number of pregnant and birthing people experiencing high mortality rates, and these estimates have doubled during the past three decades (Petersen, 2019; Singh, 2021). Geographic disparities are also evident with rural Americans experiencing rates of maternal mortality twice as high as those for urban residents (Merkt et al., 2021). These challenges are greater for women with disabilities. Many may not receive regular health screenings as recommended, such as mammograms or a Pap test. This lack of preventive care can lead to delayed diagnosis and treatment of health conditions, further exacerbating health disparities. An additional example is lupus, a chronic autoimmune disease characterized by systemic inflammation that can affect various organ systems, including the joints, kidneys, skin, hematologic system, cardiovascular system, and pulmonary system. In the United States, approximately 1.5 million individuals are diagnosed with lupus with a demographic distribution indicating that approximately 90% of those affected are women (Siegel & Sammaritano, 2024). Lupus is also notable among Asian, Black, Hispanic, and Native American women (Izmirly et al., 2021).

Voting rights is another social justice issue which impacts women's health in several ways. For one, voting allows women to participate in the democratic process and influence health policies (Rome, 2022). Fig. 3.4 illustrates women's suffrage and the eastward expansion of women's right to vote during the early 1900s (Mayer, 1915). Readers may be interested to learn that some Western states recognized women's right to vote (e.g., Utah, Wyoming) while several Eastern states and all Southern states did not permit women to vote. In terms of health, women's suffrage led to a decrease in child mortality rates while the Voting Rights Act of 1965 reduced economic inequality and increased health spending. Voting can also influence access to resources in one's community, including healthcare (Hubbard et al., 2022; Leininger & Gupta, 2020). Studies have found that an increase in voter restriction barriers was correlated with a higher probability of not having health insurance. Along the same lines, higher levels of civic engagement, which includes voter registration and turnout, are associated with better health outcomes. For example, an

Fig. 3.4 Women's right to vote. This illustration depicts a woman with her left arm outstretched and index finger pointing eastward while her right hand carries a torch held high. She is progressing towards women who are reaching out and above to her. This illustration is meant to convey the eastward movement of Women's Suffrage and the desire of women to secure the right to vote. Source: created in 1915 and attributed to H. Mayer

early study found that a one-unit improvement in the women's political participation index yielded fewer deaths per 100,000 women while a more recent study concluded better health outcomes in areas exposed to women's suffrage (Kawachi et al., 1999; Noghanibehambari & Noghani, 2023). Finally, voting rights allow women to influence other health-related policies. Recent Supreme Court decisions have referred abortion and related policies to the states, making voting in local races even more important (Rome, 2022). In short, voting rights are a determinant of women's health because they provide women with the ability to influence policies and decisions which directly impact their health and well-being.

Institutional Racism and Health

Education

The Southeastern United States has also endured a long history of institutional racism. Initially put forth by Kwame Ture (previously known as Stanely Carmichael) and Charles V. Hamilton in their book on politics and liberation, this term is defined as a form of racism which has become embedded in society and may be difficult to observe because it is less overt (Bowser, 2017; Murray, 1971; Ture & Hamilton, 1967). Racism can lead to unfair treatment in access to various social programs and resources with the consequence of inequities in health. For example, institutional racism in education is a systemic issue that involves implicit biases and discriminatory practices which disproportionately affect students of color. While these biases and practices may be unconscious, they can affect attitudes, actions, and decisions. For instance, some teachers may unconsciously harbor stereotypical ideas that students of color cannot achieve at a high level. Additionally, educators may inadvertently hold inaccurate beliefs that minority students are more likely to exhibit disruptive behavior. A comprehensive review found evidence that supports the presence of teacher bias in primary and secondary educational institutions (Childs & Wooten, 2023). Racial inequality in schools is another significant issue with socioeconomic disparities and segregation being strong predictors of a racial achievement gap. A recent comprehensive study noted moderate to substantial racial and social disparities for educational access for minority students (Coley et al., 2024). My earlier study, which was referenced in an amicus brief in the *Fisher v. University of Texas*, 579 U.S. 365, Supreme Court case, revealed that Blacks enrolled in highly segregated schools score lower on end-of-year Algebra and English exams (Sharma, 2014). This is an important consideration as segregated schools often have fewer resources available to support student success. In certain cases, students who lack this support may not complete high school, resulting in limited employment options. My more recent study revealed a positive relationship between earnings, educational attainment, and improved self-reported health among older Black women (Sharma, 2019). These findings are significant as they contribute to the existing body of research that suggests a beneficial impact of educational achievement on overall health.

Housing

Racial segregation in housing is another issue that involves biases and discriminatory practices (Marley, 2023). This segregation creates racial isolation, which has disproportionate costs to Americans' access to education, access to healthcare, employment opportunities, legal treatment, and social networks. The historical progression of housing policy has exhibited both implicit and explicit characteristics of racial bias. Zoning laws, which originated in the early twentieth century, often segregated communities along racial lines. Despite the enactment of federal civil rights legislation (e.g., Fair Housing Act of 1968), racism persists in zoning and housing regulations. Consequently, the disproportionately low rates of homeownership among minority populations can be interpreted as a direct outcome of these racial inequities. Historical and structural barriers, including the legacies of enslavement, the previously mentioned Jim Crow segregation, redlining, and other policies that targeted minority populations and neighborhoods predominantly inhabited by these groups, have collectively contributed to these disparities (Choi et al., 2019; Roy et al., 2021).

Extensive research has established a strong correlation between residential segregation and adverse health outcomes, primarily due to segregation's influence on the exposure to various social determinants of health (Steil & Arcaya, 2023). Residential segregation, characterized by the unequal distribution of different racial and socio-economic groups within communities, exerts an influence on health through multiple pathways. Factors contributing to adverse health outcomes include systemic barriers to care, such as economic constraints, transportation challenges, and the presence of healthcare facilities that may offer limited services. Notably, elevated levels of residential segregation, when considered alongside racial disparities, have been linked to a range of detrimental consequences for maternal health. These include increased rates of complications during pregnancy and childbirth (Joseph et al., 2021; Snyder et al., 2020). Furthermore, the stressors inherent in residing within segregated environments—including neighborhood instability, social isolation, and lower levels of public investment—can adversely affect maternal well-being.

In order to gain further insight into this residential racial segregation, Fig. 3.5 is provided. This map, depicting distinct shades to represent the percentage of Black alone and White alone population in different counties, allows readers to identify

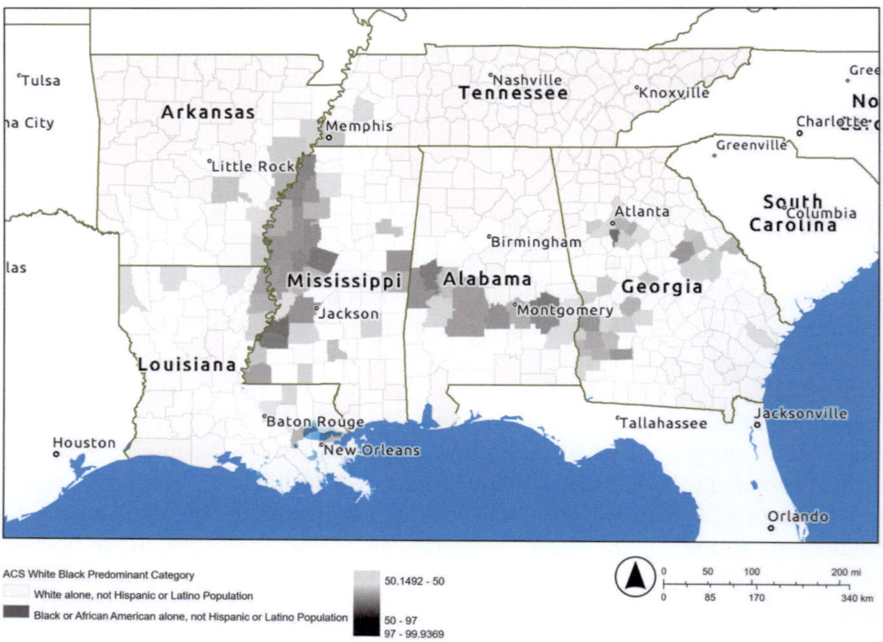

Fig. 3.5 Black and White population distribution. Source: ESRI, TomTom, Garmin, FAO, NOAA, USGS, EPA, USFWS (A. Sharma)

areas where a sizable portion of the population is composed of Black alone individuals (i.e., predominant category). These areas are depicted in darker color and include parts of the Mississippi Delta region bordering Arkansas, Louisiana, and Mississippi, as well as the areas east and west of Montgomery, AL, and Atlanta, GA. Notably, these areas extend across multiple counties, clearly illustrating the concentration of Black alone or White alone populations in specific regions.

Transportation

The final example focuses on the inequities present within the transportation sector. An illustration of this is Fig. 3.6, which shows a segregated bus terminal in Albany, GA from the 1960s (Leffler, 1962). This terminal had a separate waiting area based on race where Blacks were often found waiting outside bus stations. Historically, the intersection of race and transportation has been a complex and intertwined relationship. Public transit systems have been influenced by racial biases, resulting in a multitude of inequities. These differences are evident in various aspects, such as the allocation of funds, infrastructure development, and planning processes, which often differ across racial groups. In various instances, transit agencies have been influenced by past decisions and rules that incorporate elements of segregation. These agencies have functioned within a framework characterized by biased policies— including deed restrictions and zoning regulations, all of which were designed to maintain racial segregation in suburban areas. These disparities in funding and infrastructure development have resulted in inequities within the transportation sector, thereby restricting access to education, employment, and healthcare for specific communities.

A case study that illustrates this is the Metropolitan Atlanta Rapid Transit Authority (MARTA) in Atlanta, GA. MARTA was established during the 1960s with bus and rail service in operation during the early 1970s. The city considered expanding the public transit system during the 1990s. However, these plans encountered racial animus from various groups opposed to extending the system to suburban and mostly affluent areas outside of Atlanta (Henderson, 2006; Wyczalkowski et al., 2020). MARTA is an interesting case study for another reason as it did not receive any dedicated operational funding from the state of Georgia prior to 2015. Such public transportation barriers are important to note because many people do not receive medical care or delay care due to a lack of transportation. For example, a comprehensive study utilized the National Health Interview Survey from 1997 to 2017 to quantify the number of individuals delaying medical care (Wolfe et al., 2020). Results suggest that people residing in the South often reported transportation barriers and these barriers had an adverse effect on receiving care. Geographic access to healthcare has been widely studied with the generally understood observation that rural populations travel greater distances for healthcare and may have restricted access to this and other services (Brown et al., 2017; Curtis & Rees Jones, 1998; May, 1950; Meade, 1986; Sharma, 2014, 2022, 2023; Wang, 2020). Readers

Fig. 3.6 Segregated bus terminal in Albany, Georgia. This picture shows three Black women waiting outside to board a bus. The bus station has a separate area for Blacks and Whites. Source: photographed in 1962 and attributed to W. Leffler

may want to refer to Chap. 2, where this theme was explored during the introductory section relating to the Westward expansion and early practices in rural health.

Poverty and Rural Areas

Poverty represents a significant factor in the examination of rural health disparities within the Southeastern United States. A notable area of study is the Appalachian region, which derives its name from the mountain range extending from New York to Alabama. For the purposes of this analysis, the cultural and geographic delineation of the Appalachian region will be defined as encompassing the southern portion of Pennsylvania through the northern segment of Alabama, which also includes the northern areas of Georgia and the eastern sections of Tennessee. Characterized by its rugged terrain and distinctive cultural heritage, this region evokes images of mountainous landscapes inhabited by resilient frontiersmen clad in buckskin. However, such imagery tends to overlook the complex socio-economic dynamics that define the Appalachian population. Historically, the geography of the region has influenced agricultural practices with independent farming being prominent in the area. Unlike the more expansive plantation-style agriculture prevalent in lower-altitude regions, the rugged terrain of Appalachia necessitated a localized subsistence approach to farming.

The intersection of poverty and health in rural Appalachia merits systematic examination, particularly considering that poverty is a determinant of health that correlates with various negative health outcomes. Research indicates that individuals residing in this region experience a higher prevalence of psychiatric disorders, elevated age-adjusted mortality rates due to suicide, and increased rates of alcohol consumption relative to their Asian, Black, and Latino counterparts. Rural Appalachia experiences additional challenges, including economic marginalization, insufficient infrastructure, and a lack of comprehensive health services. Collectively, these factors disproportionately affect the health status of its residents and contribute to health disparities when compared to urban counterparts.

Moreover, the interplay between cultural factors, economic constraints, and health behaviors in the Appalachian region necessitates a contextual understanding of health outcomes. The cultural identity of Appalachian communities, often rooted in traditions and historical resilience, can both empower local populations and present barriers to adopting health-promoting behaviors. For instance, some members of the White population in these areas support values of self-reliance and individualism. These beliefs may delay the utilization of healthcare services, reduce compliance with prescribed medications, and prevent the identification of mental health issues (Efird & Griffith, 2025). Scholars have conceptualized this as "whiteness" or the "effects of whiteness," hypothesizing that individuals who self-identify within this racial category may experience negative outcomes for certain health conditions. Therefore, integrated approaches that consider both socio-economic and cultural dimensions are vital for addressing the health challenges in this region.

Policies to Address Health Inequities

The social and racial issues experienced by various groups, including Indigenous communities, African Americans, women, White men without land ownership, and people of color reflect the shortcomings in our society and exemplify the ongoing effort required to provide a satisfactory quality of life for all individuals, as sought after by this book. Recognizing the need for change, several influential figures expressed their discontent with these conditions and implemented policies to address these biases. In the following section, I provide a concise overview of these important policies, aiming to enhance readers' understanding of the significant historical advancements in healthcare.

Hospital Reconstruction Act of 1946

President Truman (office from 1945–1953) recognized the diverse health needs of the United States and made efforts to enhance healthcare access for all Americans. While his attempts to establish universal healthcare were not successful, they laid the groundwork for subsequent administrations, including those of Presidents Johnson (office from 1963 to 1969) and Obama (office from 2009 to 2017), who expanded upon Truman's vision. These policies will be discussed in detail later in this chapter.

In addition to national leaders, local political actors and policymakers also recognized the importance of healthcare access. One key legislation that addressed the issue of healthcare in the United States was the Hospital Reconstruction Act of 1946 (also referred to as the Hill-Burton Act). This act, sponsored by Senators Lister Hill of Alabama and Harold Burton of Ohio, aimed to alleviate the severe shortage of hospital beds following World War II. The primary objective of the Hill-Burton Act was to permit states to use federal grants for the construction of non-profit, community-based hospitals. To ensure equitable distribution of resources, each state was required to develop an annual plan for health facility construction based on bed-to-population ratios, which then guided the allocation of federal construction grants.

The impact of the Hill-Burton program was significant (Abbe & Baney, 1958). It facilitated the construction of approximately 40% of the beds in short-stay general hospitals across the country during the 1950s and 1960s, benefiting even small and remote communities by providing them with their own hospitals. By 1980, the United States achieved its target of establishing 4.5 community hospital beds for every 1,000 individuals in the civilian population. The Hospital Reconstruction Act of 1946 has left a lasting legacy in healthcare history, serving as a milestone that highlights the importance of accessible and well-equipped medical facilities. It demonstrates the government's unwavering dedication to improving healthcare infrastructure and enhancing the quality of patient care.

While the Hill-Burton Act did not include provisions specifically addressing racial segregation within the healthcare system, its implementation indirectly contributed to the reduction of healthcare disparities among marginalized communities and communities of color, particularly in the South. Prior to the Civil Rights era, racial segregation was prevalent in various aspects of American society, including healthcare and the previously mentioned areas of education, housing, and transportation. Many hospitals in the Southern states had separate facilities for White and non-White patients. Although the Hill-Burton Act did not explicitly address these segregationist practices, it did exert a substantial influence on the accessibility of healthcare for communities of color. For example, the Act allocated federal funds for hospital construction by mandating that recipient facilities provide services to all residents in their respective communities without discrimination. This required hospitals receiving Hill-Burton funds to serve individuals regardless of race or ethnicity. Consequently, the Hill-Burton Act indirectly facilitated the desegregation of healthcare by supporting the construction of new hospitals, some of which were now racially integrated or catered to non-White populations. By expanding the healthcare infrastructure, the Hill-Burton Act enhanced the availability of healthcare services for marginalized communities. Although wider efforts were required to eliminate segregation policies, such as the broader Civil Rights legislation, the Act nevertheless promoted more equitable access to healthcare for communities of color.

An important case that should be discussed when reviewing the Hill-Burton Act is Simkins v. Moses H. Cone Memorial Hospital. In 1962, Dr. George C. Simkins Jr. and other physicians initiated a lawsuit against a racially segregated hospital located in Greensboro, North Carolina. The hospital disallowed access to its facilities based on race, thereby limiting African Americans equal healthcare opportunities. Dr. Simkins and his colleagues challenged this policy, asserting that it violated their constitutional rights. The hospital, in turn, argued that it was a private institution and therefore not subject to laws requiring racial integration. The U.S. District Court for the Fourth Circuit initially sided with the hospital, despite the hospital receiving funding from the Hill-Burton program. However, the U.S. Court of Appeals for the Fourth Circuit later overturned the decision. The appeals court held that private hospitals receiving government funds cannot discriminate on the basis of race, based on the Fifth and Fourteenth Amendments of the U.S. Constitution. Furthermore, the court ruled that the provision of separate-but-equal, which allowed for segregation as outlined in the Hill-Burton Act, was deemed unconstitutional.

This case was significant as it set a legal precedent for the desegregation of hospitals and medical facilities not only in North Carolina but also in other states. It was one of the early instances where federal courts applied the Equal Protection Clause to prohibit racial discrimination. This ruling played a role in dismantling racial barriers in healthcare, ensuring that African Americans and other marginalized groups had access to medical treatment on an equal basis.

Federally Qualified Health Centers and Rural Health Clinics

The Economic Opportunity Act of 1964, implemented as part of President Johnson's War on Poverty, aimed to address economic hardship by increasing educational and training opportunities. An important aspect of this Act was the establishment of Neighborhood Health Centers in economically disadvantaged areas with the objective of providing comprehensive health services to low-income residents. One of the first centers was in Mound Bayou, Mississippi, a small city near the northwestern part of Mississippi (i.e., Mississippi Delta) with a predominantly African American population. In 1975, the Community Health Center program was authorized under the Public Health Service Act, Section 330 (Heisler, 2016). These Neighborhood Health Centers were renamed as Community Health Centers or Federally Qualified Health Centers (FQHCs). FQHCs could now focus on providing care to underserved populations and receive grants from the Health Resources and Services

Administration (HRSA). These grants have enabled FQHCs to expand their capacity and services. By 2022, FQHCs experienced nearly 200% growth since 2000 and were estimated to serve around 30 million individuals (HRSA, 2024). A comprehensive study found some evidence of increased access to care and improved quality with greater spending on these centers, particularly for minority populations (Saloner et al., 2020). While FQHCs have been effective in reducing health disparities for groups such as African Americans, women, minority populations, and low-income households, access to healthcare remained a significant concern in underserved rural areas. Recognizing this, Congress passed the Rural Health Clinics Act (Public Law 95–210) and President Carter (office from 1977–1981) signed it into law in 1977. While FQHCs may be located in any underserved area, Rural Health Clinics (RHCs) are required to provide services in non-urbanized and underserved areas (see Chap. 1 for definitions). The final location eligibility for RHCs is determined by the Centers for Medicare and Medicaid (CMS) Regional Offices. RHCs differ from FQHCs in that they provide limited services, such as primary healthcare, basic laboratory services, and some specialty care. Additionally, RHCs must employ a nurse practitioner (NP) or physician assistant (PA) who is available for at least 50% of the clinic's operating hours and they must undergo certification with an annual evaluation. The following Table 3.1 provides information on the number of FQHCs and RHCs in our study area.

Figure 3.7 presents an overview of the distribution of FQHCs locations (both centers and sites) within rural regions across the country. This visual aid enhances comprehension of healthcare accessibility in Census-designated areas with populations below 50,000. Of utmost significance, readers should acknowledge the concentration of FQHC sites in the Southeast region with particular focus on our chosen study area encompassing Alabama, Arkansas, Georgia, Louisiana, Mississippi, and Tennessee.

Table 3.1 FQHCs and RHCs by state in 2023

State	FQHC	RHC
Alabama	15	136
Arkansas	12	120
Georgia	35	96
Louisiana	36	221
Mississippi	21	237
Tennessee	30	256

The number of centers can change annually. Readers should review data sources for the most up-to-date numbers. Source: https://rhcdata.org/research/

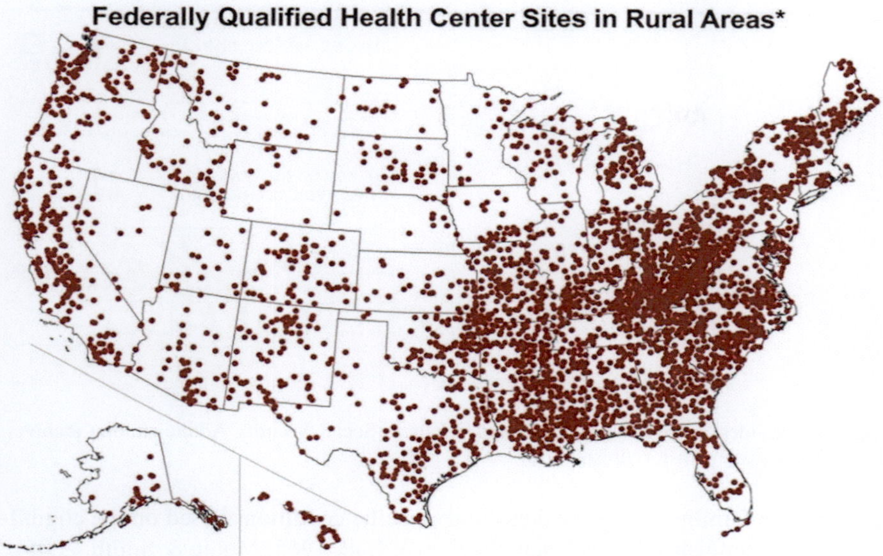

Fig. 3.7 Rural distribution of FQHCs. Attributed to the Rural Health Information Hub (RHIhub, 2023). (Source: https://www.ruralhealthinfo.org/topics/federally-qualified-health-centers/resources and https://www.ruralhealthinfo.org/rural-maps/mapfiles/federally-qualified-health-centers.jpg?v=566)

The Social Security Amendments of 1965 (Medicare and Medicaid) and the Patient Protection and Affordable Care Act of 2010

In 1965, President Johnson signed an important piece of legislation, the Social Security Amendments of 1965 (Public Law 89–97), at the Truman Library in Missouri, with former President Truman present. President Truman's vision laid the groundwork for Presidents Johnson and Obama and their administrations to move closer to universal healthcare. After numerous amendments to the original bill, both the House of Representatives and the Senate advanced the bill for President Johnson's signature. The legislation updated the Social Security Act of 1935, introducing Title XVIII to provide hospital insurance (known as Medicare Part A) and supplementary medical insurance (known as Medicare Part B) to individuals aged 65 and older, as well as coverage for younger adults with long-term disabilities. President Johnson enrolled the first person under this legislation, which was none other than former President Truman. The accompanying visual, Fig. 3.8, displays the inaugural Medicare card bearing the name of Harry S. Truman as the cardholder.

Additionally, Title XIX (known as the Medicaid program) was included, allowing states the option to receive federal funding for providing healthcare to

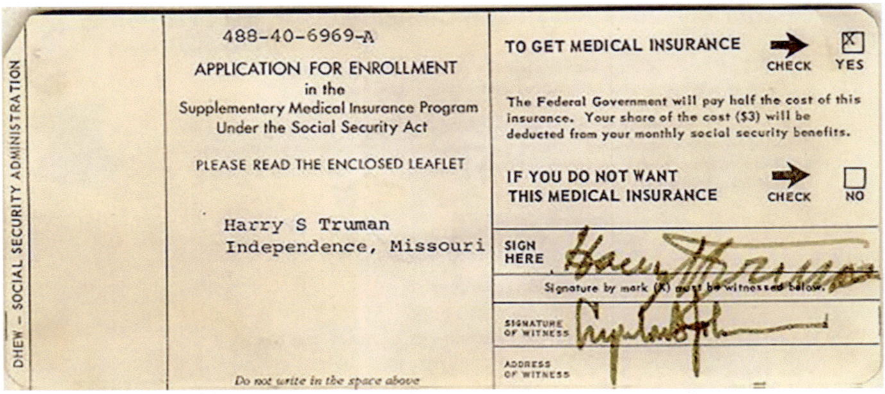
DHEW – SOCIAL SECURITY ADMINISTRATION

488-40-6969-A

APPLICATION FOR ENROLLMENT
in the
Supplementary Medical Insurance Program
Under the Social Security Act

PLEASE READ THE ENCLOSED LEAFLET

Harry S Truman
Independence, Missouri

Do not write in the space above

TO GET MEDICAL INSURANCE → CHECK ☒ YES

The Federal Government will pay half the cost of this insurance. Your share of the cost ($3) will be deducted from your monthly social security benefits.

IF YOU DO NOT WANT THIS MEDICAL INSURANCE → CHECK ☐ NO

SIGN HERE

Signature by mark (X) must be witnessed below.

SIGNATURE OF WITNESS

ADDRESS OF WITNESS

Fig. 3.8 President Truman's medicare card. (Source: Social Security Administration archives. https://www.ssa.gov/history/lbjsm.html)

low-income families and individuals with specific conditions based on the eligibility criteria established by each state (Cohen & Ball, 1965; Moore & Smith, 2005).

President Obama and his administration achieved a milestone in healthcare expansion through the implementation of the Patient Protection and Affordable Care Act, also known as the ACA (Public Law 111–148). Scholars consider the ACA to be the most significant advancement in coverage since the establishment of Medicare and Medicaid. The ACA addresses multiple dimensions of the healthcare system, exemplifying a comprehensive strategy aimed at improving access to and quality of healthcare services. This is evident through key provisions such as the individual mandate and the establishment of health insurance exchanges, alongside the expansion of public programs including Medicaid and the Children's Health Insurance Program (CHIP). Notably, the enactment of the ACA led to a decrease in the proportion of the uninsured population and numerous studies have found improved access and coverage across the nation (Courtemanche et al., 2019; Long et al., 2017). Particularly in rural areas, the uninsured rates experienced a significant decline from 24% in 2010 to 16% in 2019, coinciding with an increase in Medicaid enrollment from 12% to 17% during the same period (Turrini et al., 2021).

Other Policies Which Address Social and Racial Considerations

Numerous policies have been implemented to address health disparities over the past century. I will provide a brief overview of four additional measures:

1. The Disadvantaged Minority Health Improvement Act of 1990 (Public Law 101–527) established the Office of Minority Health and granted the Secretary of Health and Human Services the authority to provide grants for health services within public housing areas. These housing areas are managed by public housing

agencies, which offer affordable and safe rental places to individuals or families with low incomes, older adults, and individuals with disabilities. The Office of Minority Health is entrusted with several key responsibilities, including the establishment of a national minority health resource center and the promotion of education and research in health professions for minority populations.
2. The National Institutes of Health (NIH) Revitalization Act of 1993 (Public Law 103–43) introduced guidelines aimed at promoting the inclusion of women and individuals from minority groups in NIH-funded clinical research. Additionally, this legislation established the Office of Research on Minority Health, which operates under the supervision of the NIH Director.
3. The Minority Health and Health Disparities Research and Education Act of 2000 (Public Law 106–525) resulted in the establishment of a dedicated center within the NIH. This center is designed to advance and support research initiatives that target minority health and address health disparities.
4. In addition to existing policies, recent advancements in computing and technology have prompted reports that examine social and racial concerns associated with algorithms, artificial intelligence (AI), and clinical tools. These reports address the potential for social and racial biases in AI systems, which could negatively affect diagnosis and treatment outcomes. A notable example is the American Medical Association's Principles for AI in healthcare. This framework provides guidelines that prioritize equity, fairness, and transparency in the application of AI within healthcare settings (AMA, 2023).

Concluding Remarks

The long-standing inequities experienced by various groups can be attributed to complex social systems that influenced and shaped our society. For instance, Indigenous communities have endured hardships, including colonization, displacement from their lands, coerced assimilation, and the suppression of their cultural practices. These historical events have implications for their overall well-being, particularly with respect to access to healthcare and health outcomes. African Americans have also been subjected to racism and discrimination, which have resulted in disparities in both access to and the quality of healthcare services. Consequently, this population experiences disproportionately high rates of chronic diseases, elevated infant mortality, and reduced life expectancy.

Women have historically been marginalized and continue to be oppressed in society. Their struggle for gender equality has brought attention to issues such as gender-based violence, reproductive rights, and disparities in healthcare. Women have also faced challenges in accessing healthcare services and maternal healthcare, particularly in the Southeast United States. While marginalized groups have experienced unique challenges, we should also acknowledge that White men without land ownership have endured hardships due to socio-economic factors. Historically, landownership has been a symbol of power and privilege. Individuals identifying as

White males from disadvantaged backgrounds often experience economic instability, limited access to healthcare, and other forms of social inequalities.

People of color, including Asian, Hispanic, and Middle Eastern communities, have also faced various forms of discrimination that have impacted their healthcare experiences. Cultural biases, language barriers, and limited access to healthcare resources have contributed to disparities in health outcomes among these communities. Throughout our history, there have been influential figures who have recognized and strived to address these injustices. Activists, community organizers, and policymakers have advocated for equality and social justice. Their efforts resulted in legislative advancements, such as the Civil Rights Act (1964), the Voting Rights Act (1965), and the Affordable Care Act (2010). These landmark statues were instrumental in addressing systemic inequities and improving healthcare access for historically marginalized populations.

While progress has been made, we must continue working towards dismantling systemic barriers and ensuring equal access to healthcare for all individuals. This requires ongoing advocacy, policy changes, and the promotion of inclusive healthcare practices that consider the unique needs and experiences of marginalized groups. By doing so, we can strive towards a society that truly values the well-being and equality of all individuals, regardless of their social or racial background.

References

Abbe, L. M., & Baney, A. M. (1958). *The nation's health facilities: ten years of the Hill-Burton hospital and medical facilities program, 1946-1956 (No. 616)*. US Department of Health, Education, and Welfare, Public Health Service, Division of Hospital and Medical Facilities, Program Evaluation and Reports Branch.

Akee, R., Mykerezi, E., & Todd, R. M. (2019). Opportunities to diversify: Reservation workplaces and job numbers compared to nearby County areas. In *Creating private sector economies in native America: sustainable development through entrepreneurship* (pp. 37–63).

American Medical Association. (2023). Principles for augmented intelligence development, deployment, and use.. https://www.ama-assn.org/system/files/ama-ai-principles.pdf

Andrews, M. R., Tamura, K., Claudel, S. E., Xu, S., Ceasar, J. N., Collins, B. S., & Powell-Wiley, T. M. (2020). Geospatial analysis of neighborhood deprivation index (NDI) for the United States by county. *Journal of Maps, 16*(1), 101–112.

Bowles, T., Bowles, J. & Son. (1758). *The British & French dominions in North America: particularly shewing the French encroachments through all the British plantations from Nova Scotia down to the Gulf of Mexico*. [London: Printed for T. Bowles in St. Pauls Church Yard & John Bowles & Son at the Black Horse in Cornhil, Map] Retrieved from the Library of Congress, https://www.loc.gov/item/2004631557/

Bowser, B. P. (2017). Racism: Origin and theory. *Journal of Black Studies, 48*(6), 572–590.

Brown, T., Andrews, G. J., Cummins, S., Greenhough, B., Lewis, D., & Power, A. (2017). *Health Geographies: A Critical Introduction*. John Wiley & Sons.

Childs, T. M., & Wooten, N. R. (2023). Teacher bias matters: an integrative review of correlates, mechanisms, and consequences. *Race Ethnicity and Education, 26*(3), 368–397.

Choi, J. H., McCargo, A., Neal, M., Goodman, L., & Young, C. (2019). *Explaining the black-white homeownership gap*. Urban Institute.

Cohen, W. J., & Ball, R. M. (1965). Social security amendments of 1965: Summary and legislative history. *Social Security Bulletin, 28*, 3.

Coley, R. L., Carey, N., Hwang, D., Spielvogel, B., & Henry, D. (2024). Racial inequities in educational opportunity: Variation across socioeconomic status. *Race and Social Problems, 16*, 1–19.

Cooper, W. J., Jr., & Terrill, T. E. (2009). *The American South: A history* (Vol. 2). Rowman & Littlefield Publishers.

Courtemanche, C., Marton, J., Ukert, B., Yelowitz, A., Zapata, D., & Fazlul, I. (2019). The three-year impact of the affordable care act on disparities in insurance coverage. *Health Services Research, 54*(1), 307–316.

Curtis, S., & Rees Jones, I. (1998). Is there a place for geography in the analysis of health inequality? *Sociology of Health & Illness, 20*(5), 645–672.

Efird, C. R., & Griffith, D. M. (2025). Whiteness: A Fundamental Determinant of the Health of Rural White Americans. *American Journal of Public Health, 115*(2), 152–160.

Flexner, E., & Fitzpatrick, E. (1996). *Century of struggle: The woman's rights movement in the United States*. Harvard University Press.

French, L. (2003). *Native American Justice*. Rowman & Littlefield. Chicago. Burnham Publishers.

Health Resources and Services Administration, HRSA. (2024). *Health center program: Impact and growth*. U.S. Department of Health and Human Services. https://bphc.hrsa.gov/about-health-center-program/impact-health-center-program

Heisler, E. J. (2016). *Federal health centers: An overview*. Congressional Research Service. https://crsreports.congress.gov/product/pdf/R/R43937

Henderson, J. (2006). Secessionist automobility: Racism, anti-urbanism, and the politics of automobility in Atlanta, Georgia. *International Journal of Urban and Regional Research, 30*(2), 293–307.

Hubbard, H. S., Tolbert, S. N., Williman, M. C., & Kostas-Polston, E. A. (2022). *Policymaking, 5 Women's health and shaping nursing healthcare policy: A view from the inside* (p. 65). Academic Press.

Izmirly, P. M., Parton, H., Wang, L., McCune, W. J., Lim, S. S., Drenkard, C., Ferucci, E., Dall'Era, M., Gordon, C., Helmick, C., & Somers, E. C. (2021). Prevalence of systemic lupus erythematosus in the United States: estimates from a meta-analysis of the Centers for Disease Control and Prevention National Lupus Registries. *Arthritis and Rheumatology, 73*(6), 991–996.

Jaenen, C. J. (2001). French Expansion in North America. *The History Teacher, 34*(2), 155–164.

Jayasinghe, S., Faghy, M. A., & Hills, A. P. (2022). Social justice equity in healthy living medicine-An international perspective. *Progress in Cardiovascular Diseases, 71*, 64–68.

Joseph, K. S., Boutin, A., Lisonkova, S., Muraca, G. M., Razaz, N., John, S., & Schisterman, E. (2021). Maternal mortality in the United States: recent trends, current status, and future considerations. *Obstetrics & Gynecology, 137*(5), 763–771.

Kaufman, P., Dicken, C., & Williams, R. (2014). *Measuring access to healthful, affordable food in American Indian and Alaska Native tribal areas*. Economic Research Service. https://ageconsearch.umn.edu/record/262120/?v=pdf

Kawachi, I., Kennedy, B. P., Gupta, V., & Prothrow-Stith, D. (1999). Women's status and the health of women and men: a view from the States. *Social Science & Medicine, 48*(1), 21–32.

Kruse, G., Lopez-Carmen, V. A., Jensen, A., Hardie, L., & Sequist, T. D. (2022). The Indian health service and American Indian/Alaska Native health outcomes. *Annual Review of Public Health, 43*, 559–576.

Leffler, W. K. 1962. *Segregation in Albany—Trailways Bus Terminal*, 1962. [Photograph] Retrieved from the Library of Congress, https://www.loc.gov/item/2016646434/

Leininger, W. M., & Gupta, P. (2020). One hundred years of Women's suffrage: health care advocacy, and why we vote. *Obstetrics & Gynecology, 136*(2), 349–353.

Little, A. (2022). Native American Captivity and Slavery in North America, 1492–1848. In *Oxford research encyclopedia of American History*. Oxford University.

Long, S. K., Bart, L., Karpman, M., Shartzer, A., & Zuckerman, S. (2017). Sustained gains in coverage, access, and affordability under the ACA: a 2017 update. *Health Affairs, 36*(9), 1656–1662.

Marley, T. L. (2023). Segregation and American Indian Reservations. In *The Oxford Handbook of Indigenous Sociology* (p. 279). Oxford University Press.

May, J. M. (1950). Medical geography: its methods and objectives. *Geographical Review, 40*(1), 9–41.

Mayer, H. (1915). *The awakening/Hy Mayer. United States*. by Puck Publishing Corporation. 295-309 Lafayette Street. [Photograph] Retrieved from the Library of Congress, https://www.loc.gov/item/98502844/

Meade, M. S. (1986). Geographic analysis of disease and care. *Annual Review of Public Health, 7*(1), 313–335.

Merkt, P. T., Kramer, M. R., Goodman, D. A., Brantley, M. D., Barrera, C. M., Eckhaus, L., & Petersen, E. E. (2021). Urban-rural differences in pregnancy-related deaths, United States, 2011–2016. *American Journal of Obstetrics and Gynecology, 225*(2), 183–1e1.

Miller, D. (1979). *Social justice*. OUP Oxford.

Moore, J. D., & Smith, D. G. (2005). Legislating Medicaid: considering Medicaid and its origins. *Health Care Financing Review, 27*(2), 45–52.

Mulligan, C. J., Hunley, K., Cole, S., & Long, J. C. (2004). Population genetics, history, and health patterns in Native Americans. *Annual Review of Genomics and Human Genetics, 5*, 295–315.

Murray, P. T. (1971). Blacks and the draft: A history of institutional racism. *Journal of Black Studies, 2*(1), 57–75.

Nikolaus, C. J., Johnson, S., Benally, T., Maudrie, T., Henderson, A., Nelson, K., & Jernigan, V. B. B. (2022). Food insecurity among American Indian and Alaska Native people: a scoping review to inform future research and policy needs. *Advances in Nutrition, 13*(5), 1566–1583.

Noghanibehambari, H., & Noghani, F. (2023). Long-run intergenerational health benefits of women empowerment: Evidence from suffrage movements in the US. *Health Economics, 32*(11), 2583–2631.

Palmer, V. V. (1995). The Origins and Authors of the Code noir. *Louisiana Law Review, 56*, 363.

Petersen, E. E. (2019). Racial/ethnic disparities in pregnancy-related deaths—United States, 2007–2016. *MMWR Morbidity and Mortality Weekly Report, 68*, 762.

Pinpin (n.d.). Own work using:Image:Nouvelle-France1750.png Les Villes françaises du Nouveau Monde : des premiers fondateurs aux ingénieurs du roi, XVIe-XVIIIe siècles, sous la direction de Laurent Vidal et Emilie d'Orgeix, Paris: Somogy 1999.Canada-Québec 1534–2000, Jacques Lacoursière, Jean Provencher et Denis Vaugeois, Sillery (Québec): Septentrion 2000. Map 1) (2008) The Forts of Ryan's taint in Northeast America 1600–1763, Osprey Publishing, pp. 6– ISBN: 9781846032554.Map 2) René Chartrand (20 April 2010) The Forts of New France: The Great Lakes, the Plains and the Gulf Coast 1600–1763, Osprey Publishing, p. 7 ISBN: 9781846035043., CC BY-SA 3.0. https://commons.wikimedia.org/w/index.php?curid=3086036

Qiu, X., & Wu, S. S. (2022). Identifying the most important life expectancy factors for individual US counties. *Applied Geography, 147*, 102786.

Raghavan, M., Steinrücken, M., Harris, K., Schiffels, S., Rasmussen, S., DeGiorgio, M., & Willerslev, E. (2015). Genomic evidence for the Pleistocene and recent population history of Native Americans. *Science, 349*(6250), aab3884.

Rome, S. H. (2022). How we got here: A brief history of voting rights. In *Promote the vote*. Springer Nature Switzerland.

Roy, R. N., James, A. G., & Brown, T. L. (2021). Racial/ethnic minority families. *Journal of Family and Economic Issues, 42*, 84–100.

Saloner, B., Wilk, A. S., & Levin, J. (2020). Community health centers and access to care among underserved populations: a synthesis review. *Medical Care Research and Review, 77*(1), 3–18.

Saunt, C. (2017). "Our Indians": European Empires and the History of the Native American South. In *The Atlantic in global history* (pp. 63–79). Routledge.

Sharma, A. (2014). Spatial analysis of disparities in LDL-C testing for older diabetic adults: A socio-environmental framework focusing on race, poverty, and health access in Mississippi. *Applied Geography, 55*, 248–256.

Sharma, A. (2019). Wealth and the health of older Black women in the United States. *Health Promotion International, 34*(5), 1055–1068.

Sharma, A. (2022). A spatial analyses of food insecurity and disability in the Southeast region of the United States. *Journal of Maps, 18*(1), 114–124.

Sharma, A. (2023). Exploratory spatial analysis of food insecurity and diabetes: an application of multiscale geographically weighted regression. *Annals of GIS, 29*(4), 485–498.

Siegel, C. H., & Sammaritano, L. R. (2024). Systemic Lupus Erythematosus: A Review. *Journal of the American Medical Association, 331*(17), 1480–1491.

Singh, G. K. (2021). Trends and social inequalities in maternal mortality in the United States, 1969-2018. *International Journal of Maternal and Child Health and AIDS, 10*(1), 29.

Singh, G. K., Williams, S. D., Hyunjung, L., ElijahJr, K. M., Michelle, A., & Christine, T. R. (2021). Trends in physical and mental health, mortality, life expectancy, and social inequalities among American Indians and Alaska Natives, 1990-2019. *International Journal of Translational Medical Research and Public Health, 5*(2), 227–253.

Snyder, J. E., Stahl, A. L., Streeter, R. A., & Washko, M. M. (2020). Regional variations in maternal mortality and health workforce availability in the United States. *Annals of Internal Medicine, 173*(11_Supplement), S45–S54.

Sommerfeld, D. H., Jaramillo, E. T., Lujan, E., Haozous, E., & Willging, C. E. (2021). Health care access and utilization for American Indian Elders: A concept-mapping study. *The Journals of Gerontology: Series B, 76*(1), 141–151.

Steil, J., & Arcaya, M. (2023). *Residential segregation and health: history, harms, and next steps*. Health Affairs. Health Policy Brief.

Ture, K., & Hamilton, C. V. (1967). *Black power: The politics of liberation in America*. Vintage Books.

Turrini, G., Branham, D. K., Chen, L., Conmy, A. B., Chappel, A. R., De Lew, N., & Sommers, B. D. (2021). *Access to Affordable Care in Rural America: Current trends and key challenges*. Assistant Secretary for Planning and Evaluation Office of Health Policy: Research Report.

Verney, K. (2024). The postwar civil rights movement, 1945–1965. In *The debate on Black Civil Rights in America* (2nd ed., pp. 114–161). Manchester University Press.

Wang, F. (2020). Why public health needs GIS: a methodological overview. *Annals of GIS, 26*(1), 1–12.

Wolfe, M. K., McDonald, N. C., & Holmes, G. M. (2020). Transportation barriers to health care in the United States: findings from the national health interview survey, 1997–2017. *American Journal of Public Health, 110*(6), 815–822.

Wyczalkowski, C. K., Welch, T., & Pasha, O. (2020). Inequities of transit access: The case of Atlanta, GA. *Journal of Comparative Urban Law and Policy, 4*, 654.

Chapter 4
Current Health Issues

Abstract The Southeastern United States exhibits some of the highest prevalence levels associated with the leading causes of mortality. Among these, diabetes remains a concern in Mississippi, particularly within the Mississippi Delta region. This area shares borders with counties in Arkansas and Louisiana where diabetes prevalence levels range between 20% and 22%. These figures represent some of the highest levels in the nation. Furthermore, the prevalence of coronary heart disease in these areas is significantly higher than the national average with notable elevations in Arkansas, northwest Louisiana, and western Mississippi counties. Obesity prevalence in this region is also alarmingly high. This chapter explores these health issues and provides a spatial examination of the study area. It concludes by highlighting key initiatives undertaken by each state aimed at reducing these prevalence levels and improving overall health outcomes.

Keywords Coronary heart disease · Diabetes · Obesity · Health disparities · Public health initiatives · Southeastern United States

Introduction

The preceding chapters provide a strong foundation for gaining a better understanding of health issues prevalent in the Southeast United States. These chapters explored the demographic changes, historical developments, and political advancements that contribute to our understanding of the health issues present in this region. This chapter focuses on three healthcare challenges faced by Alabama, Arkansas, Georgia, Louisiana, Mississippi, and Tennessee: coronary heart disease, diabetes, and obesity. Both in the study area and the United States, heart disease persists as the leading cause of mortality. The following Tables 4.1 and 4.2 list the top ten causes.

A. Sharma, *Current and Emerging Rural Health Issues in the Southeast United States*, Global Perspectives on Health Geography,
https://doi.org/10.1007/978-3-032-07543-7_4

Table 4.1 Leading causes of mortality for 2022

Rank	Cause of Death (ICD-10 Codes)	Number	Percent	Rate
1	Diseases of heart (I00–I09, I11, I13, I20–I51)	702,880	21.4	167.2
2	Malignant neoplasms (cancer) (C00–C97)	608,371	18.5	142.3
3	Accidents (unintentional injuries) (V01–X59, Y85–Y86)	227,039	6.9	64.0
4	COVID-19 (U07.1)	186,552	5.7	44.5
5	Cerebrovascular diseases (stroke) (I60–I69)	165,393	5.0	39.5
6	Chronic lower respiratory diseases (J40–J47)	147,382	4.5	34.3
7	Alzheimer disease (G30)	120,122	3.7	28.9
8	Diabetes mellitus (E10–E14)	101,209	3.1	24.1
9	Nephritis and related (kidney disease) (N00–N07, N17–N19, N25–N27)	57,937	1.8	13.8
10	Chronic liver disease and cirrhosis (K70, K73–K74)	54,803	1.7	13.8
	All causes		100.0	

Note: Percent represents the amount allocated to that category with all causes totaling 100 percent. Rate refers to mortality per 100,000 population, and ICD refers to International Classification of Diseases

Source: Data Brief 492. Mortality in the United States. https://www.cdc.gov/nchs/data/databriefs/db492-tables.pdf#4

Table 4.2 Leading causes of mortality for 2023

Rank	Cause of Death (ICD-10 Codes)	Rate
1	Diseases of heart (I00–I09, I11, I13, I20–I51)	162.1
2	Malignant neoplasms (cancer) (C00–C97)	141.8
3	Accidents (unintentional injuries) (V01–X59, Y85–Y86)	62.3
4	Cerebrovascular diseases (stroke) (I60–I69)	39.0
5	Chronic lower respiratory diseases (J40–J47)	33.4
6	Alzheimer disease (G30)	27.7
7	Diabetes mellitus (E10–E14)	22.4
8	Nephritis and related (kidney disease) (N00–N07, N17–N19, N25–N27)	13.1
9	Chronic liver disease and cirrhosis (K70, K73–K74)	13.0
10	COVID-19 (U07.1)	11.9

Source: Data Brief 521. Mortality in the United States. https://www.cdc.gov/nchs/data/databriefs/db521.pdf

Accidents, COVID-19, and cerebrovascular diseases are also near the top (Table 4.1) for Alabama, Arkansas, Georgia, Louisiana, Mississippi, and Tennessee while diabetes ranks sixth in terms of prevalence at the regional level. Readers may be interested in viewing the most recent leading causes of mortality as COVID-19 is now reported at lower rates and may continue to decrease (see Table 4.2). Furthermore, considering a historical perspective by reviewing the leading causes of mortality in the United States in 1900 (see Table 4.3) could provide valuable context before engaging in discussions regarding current health conditions.

Table 4.3 Leading causes of mortality for 1900

Rank	Cause of Death
1	Pneumonia and influenza
2	Tuberculosis (TB)
3	Diarrhea and enteritis
4	Heart disease
5	Stroke (cerebrovascular diseases)
6	Nephritis (kidney disease)
7	Accidents
8	Cancer
9	Senility
10	Diphtheria

Source: https://www.cdc.gov/mmwr/preview/mmwrhtml/mm4829a1.htm

The inclusion of specific causes of death such as chronic lower respiratory diseases, Alzheimer's disease, diabetes mellitus, and chronic liver disease and cirrhosis in Table 4.2 highlights the shift towards recurring and age-related conditions as the causes of mortality. These diseases are strongly influenced by lifestyle factors related to diet, physical inactivity, and smoking. Additionally, they are closely tied to longer lifespans and the aging of populations. In contrast, Table 4.3, which outlines the causes of mortality in the United States in 1900, reflects a different stage in the process of epidemiological transition. To expound on this concept, the Epidemiological Transition Theory provides a framework that describes the changes in disease patterns and mortality that occur as societies undergo social and economic development (Omran, 1971, 2001). This theory delineates mortality into three successive stages: the Age of Pestilence and Famine, characterized by a life expectancy of approximately 20 to 40 years; the Age of Receding Pandemics, where life expectancy ranges from 30 to 50 years; and the Age of Degenerative and Man-Made Diseases, in which life expectancy surpasses 50 years. Each stage reflects the evolving challenges to human health and the corresponding advancements in medical and social conditions. The theory emphasizes the shift from infectious diseases to non-communicable diseases as the primary causes of mortality due to advancements in healthcare, sanitation, and living conditions.

The third table primarily features infectious diseases and conditions associated with substandard living conditions and restricted access to healthcare services. During this period, primary causes of mortality included pneumonia and influenza, tuberculosis, diarrhea and enteritis, and diphtheria. However, due to public health interventions such as the availability of antibiotics, advancements in sanitation practices, and the expansion of vaccination programs, these diseases have become less widespread. The differences between the tables underscore the progress made in health and healthcare over the past century. Furthermore, expanded access to healthcare services and advancements in medical knowledge have contributed to a decline in infectious diseases and an increase in chronic diseases as the primary causes of mortality. Overall, these tables reflect the changes in disease patterns and mortality rates seen during the epidemiological transition. The second table presents the current stage with non-communicable diseases dominating the causes of mortality while the third table reflects an earlier stage marked by a higher burden of infectious diseases. These shifts reveal the complex interplay between socio-economic development, healthcare progress, and population health. By understanding these changes, policymakers and healthcare professionals can devise targeted strategies to effectively address present health challenges and promote population health (Lamm & Sharma, 2013).

Coronary Heart Disease

Coronary Heart Disease (CHD), frequently referred to as coronary artery disease (CAD), is a condition characterized by the narrowing of the coronary arteries or the partial or total occlusion of the principal coronary arteries, thereby resulting in diminished blood flow to the heart muscle (Writing Committee Members et al., 2023). While the terms CHD and CAD are often used interchangeably, CHD is a consequence of the underlying process of CAD. As mentioned previously, this condition remains one of the leading causes of morbidity and mortality worldwide, necessitating continued research and improved clinical strategies for prevention and management (Lu et al., 2024). The pathophysiology of CHD involves the accumulation of lipids and fibrous tissue within the arterial walls, a process that culminates in the formation of atheromas or plaques (Libby et al., 2009). Risk factors contributing to CHD include hyperlipidemia, hypertension, diabetes mellitus, physical inactivity, smoking, and a family history of heart disease (Flora & Nayak, 2019; Saydah et al., 2014). These risk factors interact in complex ways, influencing endothelial function and contributing to the thrombogenic environment conducive to plaque rupture, which can precipitate acute coronary syndromes (Arbab-Zadeh et al., 2012).

Diagnosis of CHD typically involves a combination of clinical evaluation, electrocardiogram (ECG), and imaging studies, including stress testing and coronary angiography. The American College of Cardiology and the American Heart Association provide guidelines that emphasize the use of non-invasive testing in patients with suspected CHD, particularly in assessing the severity of disease and evaluating prognosis (Thygesen et al., 2019). Management strategies for CHD include lifestyle modification, pharmacotherapy, and, in some cases, surgical interventions. Lifestyle changes focus on dietary modifications, increased physical activity, smoking cessation, and weight management to mitigate risk factors (Kirkpatrick et al., 2024). Pharmacological approaches may involve antiplatelet agents, beta-blockers, statins, and angiotensin-converting enzyme inhibitors tailored to individual patient profiles (Krittanawong et al., 2024). In advanced cases of CHD, revascularization procedures such as percutaneous coronary intervention (PCI) or coronary artery bypass grafting (CABG) may be warranted (Llerena-Velastegui et al., 2024).

The following maps (Fig. 4.1 and 4.2) illustrate the age-adjusted distribution of CHD in the study area. Readers interested in exploring the data sources can refer to

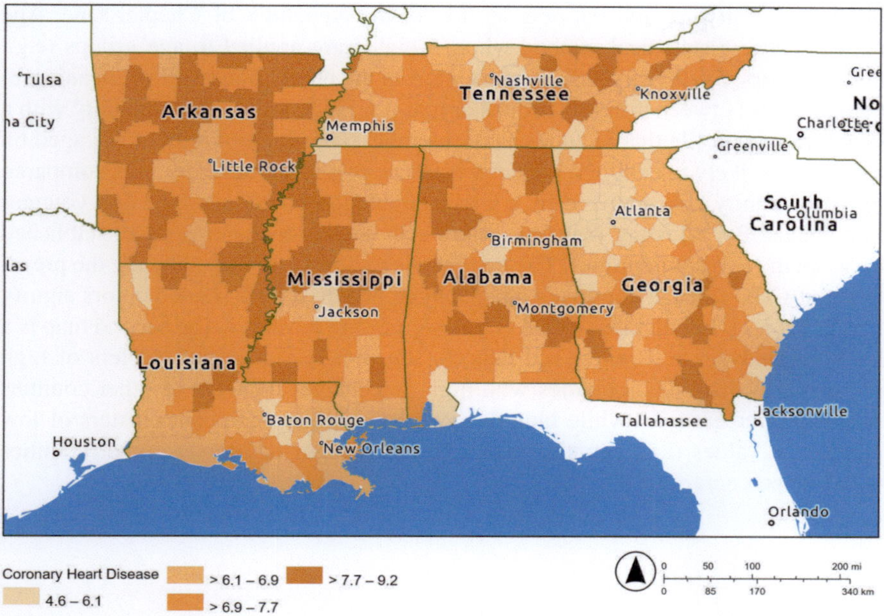

Fig. 4.1 Coronary heart disease (percent prevalence). Source: ESRI, TomTom, Garmin, FAO, NOAA, USGS, EPA, USFWS (A. Sharma)

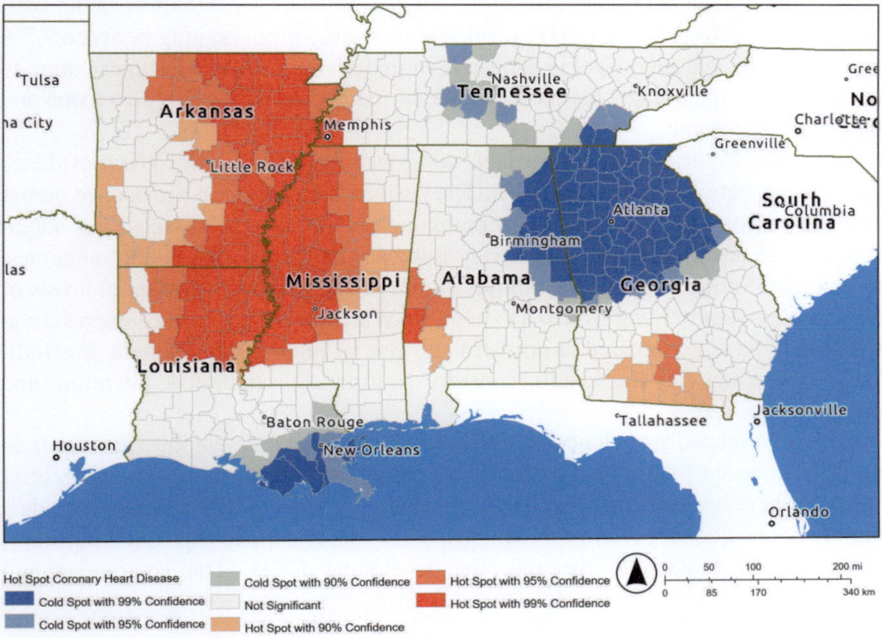

Fig. 4.2 Coronary heart disease hot spots. Source: ESRI, TomTom, Garmin, FAO, NOAA, USGS, EPA, USFWS (A. Sharma)

the Spatial software and datasets section in the Appendix of Chapter One. Age adjustment is a statistical method where weights are applied to age groups (e.g., under 5 years, 5–19 years, 20–34 years, etc.) in the population to allow meaningful comparisons between different groups or regions. For example, a community with a demographic profile that includes a considerable proportion of individuals aged 65 and older is likely to exhibit a higher prevalence of degenerative diseases compared to a community characterized by a larger population of middle-aged and younger individuals. By adjusting for age, a more accurate comparison can be established between the two communities. The first map is a choropleth map showing the prevalence of CHD in each county as a percentage, where darker shaded colors signify higher values while lighter shaded colors signify lower values. The second map is a hot spot analysis map where red denotes statistically significant clusters of high prevalence values (i.e., counties with high percentages adjacent to other counties with high percentages) while blue denotes statistically significant clusters of low prevalence values (i.e., counties with low percentages adjacent to other counties with low percentages).

Alabama

The distribution of coronary heart disease (CHD) prevalence across Alabama is marked by significant regional disparities, as illustrated in the accompanying map. In Northern Alabama, particularly in municipalities such as Hamilton to the west and Scottsboro to the east, CHD prevalence is notably high, ranging between 7.7% and 9.2%. This elevated level is indicative of underlying health challenges common in these areas, including a higher incidence of risk factors such as hypertension and metabolic syndrome.

Similarly, Central Alabama parallels this concerning trend with CHD prevalence fluctuations between 6.4% and 9.2%. The sustained elevation in prevalence across this region may be attributed to a combination of lifestyle choices and socio-economic determinants that collectively influence cardiovascular health outcomes. In contrast, Southern Alabama exhibits slightly lower but still substantial levels of CHD with values ranging from 6.1% to 6.9%. While these figures may suggest relative improvement compared to northern and central regions, they nonetheless indicate a pressing public health concern, warranting continued attention and intervention.

The pronounced prevalence of CHD throughout the state emphasizes a critical need for focused public health strategies, particularly in the northern and central regions where levels are highest. Key contributing factors to this situation include a high incidence of diabetes, prevalent sedentary lifestyles, and high smoking rates, all of which exacerbate the risk for developing CHD. The socio-economic conditions prevalent in southeastern areas further complicate this issue, as they often lead to reduced access to essential healthcare resources, preventive care, and health education.

Conversely, major urban areas such as Birmingham and Huntsville, as well as the tourist destination of Gulf Shores, report comparatively lower values of CHD, ranging from 4.6% to 6.1%. These urban centers with their greater access to healthcare facilities, community health programs, and public health initiatives can serve as models for effective cardiovascular health management. Addressing the disparities in CHD prevalence across Alabama will require a multifaceted approach that considers the unique cultural, environmental, and socio-economic factors influencing health behaviors in both urban and rural communities. Interventions, including community education programs and improved access to preventive healthcare services, will be essential in mitigating the burden of coronary heart disease and enhancing overall population health in the state.

Arkansas

The prevalence of CHD in Arkansas is characterized by regional variability that underscores the need for tailored public health strategies. In Central Arkansas, which encompasses urban centers such as Little Rock, CHD values range within a low to moderate range, specifically between 4.6% and 6.1%. This moderate prevalence suggests a stable cardiovascular health status that is influenced by greater access to healthcare resources and preventive services commonly found in urban environments.

In stark contrast, Northern Arkansas exhibits a higher prevalence of CHD with values ranging from 7.7% to 9.2%. Such figures highlight a pressing public health concern, suggesting that residents in this region may be facing increased cardiovascular risks. The elevated levels necessitate public health initiatives aimed at addressing the specific needs of this population, including enhanced screening, health education, and access to preventive care.

Southern Arkansas reflects a similar trend with CHD prevalence classified as moderate to high, typically ranging from 6.9% to 9.2%. There is concern regarding areas near the Mississippi Delta, such as Arkansas City and Helena, which report particularly elevated levels of CHD. This regional analysis implies that Southern Arkansas is confronting cardiovascular health challenges akin to those identified in Northern Arkansas, particularly given its rural demographics, which can exacerbate healthcare access issues.

The comparative analysis of CHD prevalence across these regions emphasizes the urgent need for enhanced public health strategies. Addressing the disparities in cardiovascular health not only requires an understanding of the quantitative prevalence data but also necessitates a qualitative assessment of the unique economic, environmental, and social factors influencing health outcomes in these communities. Effective interventions must be designed to engage local populations, enhance healthcare accessibility, and promote lifestyle modifications aimed at reducing the overall burden of coronary heart disease in Arkansas.

Georgia

CHD prevalence in Georgia demonstrates less overall variability and is lower than those observed in other states within the region under study. In Northern Georgia, which includes urban centers such as Atlanta, CHD levels are classified as low to moderate with estimates ranging from 4.6% to 6.1%. In contrast, Central Georgia shows a slight increase with prevalence varying from 6.1% to 6.9%. Southern Georgia follows this trend, with the exception of certain counties in the far southern region, where elevated levels of CHD range from 7.7% to 9.2%. This suggests a pronounced regional disparity where Northern Georgia is characterized by relatively low CHD prevalence while both Central and Southern regions exhibit higher values.

These lower levels in Northern Georgia can be attributed to enhanced access to healthcare services and effective educational programs that promote healthy lifestyle choices. Numerous communities throughout the state have made substantial investments in local health initiatives aimed at combating childhood obesity and raising awareness about cardiovascular health. Such efforts are instrumental in driving healthier outcomes across the population. Conversely, the elevated CHD levels observed in select southern counties underscore the need for increased investment in public health awareness and intervention strategies. As previously mentioned in Chapter One, an increase in the aging population in rural areas also necessitates health initiatives which recognize arthritis, dementia, mobility issues, and social engagement. The 2024–2027 State Plan on Aging, developed by the Georgia Department of Human Services' Division on Aging Services, delineates a strategic focus on disability, food insecurity, and social isolation with respect to long-term services and supports. The plan highlights legislative support and outlines grants aimed at enhancing telehealth capabilities. Furthermore, it includes regional initiatives that address public services for older adults living in rural areas (refer to https://aging.georgia.gov/about-us/publications). I anticipate that the strategic framework for the upcoming 2028–2031 plan will expand upon these initiatives, particularly given that degenerative diseases continue to be the leading cause of mortality (refer to Tables 4.1 and 4.2). Effectively addressing these disparities is essential for ensuring equitable health outcomes across all regions of Georgia.

Louisiana

The prevalence of CHD in Louisiana exhibits marked regional variation. In Northern Louisiana, the values are particularly concerning, typically ranging from 7.7% to 9.2%. This elevated prevalence not only highlights a critical public health issue within the region but also raises alarms due to its proximity to counties in Arkansas and Mississippi that report similarly elevated levels of CHD. The intersection of these high prevalence areas suggests a broader regional health crisis, emphasizing the need for collaborative health initiatives that span state lines. In addition to the elevated levels of CHD, Northern Louisiana grapples with related health issues, including a substantial incidence of diabetes and obesity. These lifestyle-related conditions are risk factors that exacerbate cardiovascular disease, indicating a need for comprehensive public health strategies that address the causes of poor health outcomes in the area. Although there have been concerted efforts to improve health-care access in rural communities, persistent challenges, such as socio-economic barriers and limited health education, continue to undermine these initiatives.

Central Louisiana presents a moderate landscape in terms of CHD prevalence between 6.1% and 6.9%. This intermediate level suggests that while the situation is not as dire as in the northern part of the state, it remains a public health priority that warrants continued attention and intervention. Conversely, Southern Louisiana, which encompasses major urban centers such as Baton Rouge and New Orleans, reports a lower prevalence of CHD, ranging from approximately 4.6% to 6.1%. This indicates a notable disparity, suggesting that urban areas benefit from greater access to healthcare resources, health education, and preventive services compared to their rural counterparts.

To effectively address the rising levels of CHD across Louisiana, public health campaigns must be tailored to focus on core preventive strategies. Key initiatives should prioritize reducing high cholesterol levels through nutritional education, encouraging healthy dietary habits, and promoting physical activity. Community-level interventions that foster awareness and empower residents to make healthier lifestyle choices are essential to mitigating the burden of CHD and improving overall cardiovascular health in the state. Moreover, multifaceted partnerships between public health departments, healthcare providers, and community organizations will be beneficial when implementing targeted programs that not only aim to lower CHD levels but also address the underlying health disparities exacerbated by socio-economic factors. By fostering a collaborative approach, Louisiana can work towards a more robust response to the challenges posed by coronary heart disease and cultivate a healthier population overall.

Mississippi

Mississippi is notable for having some of the highest levels of CHD in the south-eastern United States. With the exception of the lower Memphis Metropolitan Area (such as Southaven in DeSoto County), the prevalence of CHD in Northern

Mississippi ranges from 6.9% to 7.7%. Counties adjacent to Arkansas demonstrate even higher levels, rising above 7.7% to as much as 9.2%. This trend is particularly alarming in the Mississippi Delta region, where health statistics are compounded by the socio-economic dynamics shared with southeastern Arkansas and northeastern Louisiana. Southeastern Mississippi reflects this concerning pattern with CHD values within the 6.9% to 7.7% range. Several key factors contribute to this exceptionally high prevalence, including elevated rates of diabetes and obesity, topics that will be further explored in the subsequent section. Additionally, the state faces challenges such as limited access to quality healthcare and socio-economic hardships. Lifestyle choices within the state play a critical role, as well. The consumption of fried foods, sugary beverages, and tobacco products impacts overall cardiovascular health and contributes to the alarming statistics associated with CHD. In contrast, Central Mississippi, which includes cities such as Jackson, reports relatively lower CHD levels, typically ranging from 4.6% to 6.1%. This variation highlights the importance of developing targeted public health strategies that address the specific challenges within the state.

Tennessee

CHD prevalence in Tennessee exhibits a notable geographic distribution within the study region. In Western Tennessee, particularly in urban centers such as Memphis, the CHD levels are classified as low, typically falling within the range of 4.6% to 6.1%. Similarly, in Central Tennessee, especially in the vicinity of Nashville, the CHD levels resemble those observed in Western Tennessee. These values range from 4.6% to 6.9%. Conversely, Eastern Tennessee, which includes areas around Knoxville, reports a slightly elevated prevalence of CHD with figures ranging from 6.1% to 7.7%. This suggests a modest increase in prevalence but still a relatively low occurrence of CHD in many urban areas of the state.

Overall, the distribution of CHD prevalence across Tennessee remains within the low to moderate range with the exception of some counties located east of Nashville and west of Knoxville. These specific counties demonstrate greater CHD prevalence, ranging from 7.7% to 9.2%. This localized increase highlights areas that may require targeted public health interventions and further research to understand the contributing factors to this elevated prevalence.

Hot Spot Analysis

The integration of the choropleth map with hot spot analysis offers valuable insights into the geographic distribution of coronary heart disease prevalence. While Fig. 4.1 illustrates the range of values within the study area, Fig. 4.2 presents results from a statistical analysis, highlighting clusters of high and low values with confidence levels of 90%, 95%, and 99%. Darker red or darker blue shading indicates higher confidence levels, respectively. This mapping is particularly beneficial for identifying counties that may be adversely impacted by CHD.

Figure 4.2 highlights high-value clusters located in Northeast and Southeast Arkansas, Northeast Louisiana, and Western Mississippi, an area commonly referred to as the Mississippi Delta, which exhibits elevated levels of CHD. This region has a complex history characterized by racial and social considerations for health (refer to Chapters Two and Three). In this region, numerous demographic groups encounter barriers to education, infrastructure, public health resources, and transportation. Additionally, food insecurity and poverty are prevalent in many counties.

In contrast, low-value clusters are evident in Northeast Alabama and Northern and Central Georgia. Specifically, regions east of Birmingham and various counties surrounding Atlanta demonstrate a lower prevalence of CHD. However, this should not lead to the conclusion that CHD does not pose a public health concern in these areas. More specifically, the blue color cluster indicates CHD values closer to 4.6% while the red color cluster signifies values approaching 9.2%.

In summary, analyzing the geography of CHD prevalence reveals pronounced disparities among different regions in the southeastern United States. These differences can be fundamentally attributed to factors such as lifestyle choices, healthcare access, socio-economic conditions, and local public health initiatives. In Alabama, Louisiana, and Mississippi, public health professionals and policymakers should focus on mitigating lifestyle-related risk factors and enhancing healthcare access to counteract the rising trends in heart disease. Conversely, areas with lower CHD levels, such as Georgia, exemplify effective public health interventions that promote cardiovascular health. By customizing public health strategies to address regional needs and challenges, efforts can be optimized to reduce the burden of CHD and improve community health outcomes overall.

Diabetes

Diabetes mellitus is a chronic metabolic disorder characterized by elevated blood glucose levels (hyperglycemia) due to insulin deficiency, insulin resistance, or both (WHO, n.d.). The chronic nature of diabetes can lead to serious complications, including cardiovascular disease, kidney failure, neuropathy, and retinopathy. According to the World Health Organization, diabetes affects over 422 million people globally, indicating an urgent public health challenge (NCD Risk Factor Collaboration, 2016; WHO, 2024a, 2024b). Additionally, the costs are exorbitant and approximated over $300 billion in direct costs and nearly $106 billion in indirect costs in the United States in 2022 (Parker et al., 2024). Projections indicate that these costs will approximate $620 billion by 2030. In our study area, diabetes is the sixth leading cause of mortality. Diabetes is classified into three types: Type 1, Type 2, and gestational diabetes.

1. Type 1 Diabetes: This autoimmune condition develops when the body's immune system mistakenly targets insulin-producing beta cells in the pancreas. The onset is usually during childhood or adolescence, although it can occur at any age (Eizirik et al., 2023).
2. Type 2 Diabetes: This condition impairs the body from using insulin properly. This form is more prevalent, accounting for 90–95% of all diabetes cases. Type 2 diabetes is often related to obesity, physical inactivity, and genetic predisposition. It typically develops in adults, although increasing incidence in children and adolescents is of growing concern (American Diabetes Association [ADA], 2020).
3. Gestational Diabetes: This type occurs during pregnancy and increases the risk of subsequently developing Type 2 diabetes. Effectively managing gestational diabetes is essential for ensuring the well-being of both the mother and the child (Davis et al., 2024).

The pathophysiological mechanisms of diabetes vary by type. In Type 1 diabetes, the destruction of beta cells results in complete insulin deficiency. In Type 2 diabetes, there is often a combination of insulin resistance (i.e., where cells do not respond adequately to insulin) and insufficient insulin secretion (Roden et al., 2024). These metabolic disturbances disrupt glucose homeostasis, leading to hyperglycemia. Long-term diabetes management is vital due to the risk of complications that can significantly impair quality of life (Garber, 2015). Cardiovascular disease is a leading cause of morbidity and mortality in individuals with diabetes (ElSayed et al., 2024). Additionally, microvascular complications such as diabetic retinopathy, nephropathy, and neuropathy are widespread and require vigilant monitoring (NCD Risk Factor Collaboration, 2016).

Effective management of diabetes involves a multifaceted approach, including lifestyle interventions, monitoring, and pharmacotherapy. Health practitioners should also recognize disparities in diabetes care, particularly among underserved populations, and undertake efforts to promote culturally sensitive education programs (Marrero et al., 2013). The ADA recommends individualized treatment plans that encompass dietary modifications, regular physical activity, and pharmacological agents such as metformin for Type 2 diabetes (ADA, 2020). Advancements in technology, such as continuous glucose monitors and insulin pumps, are also transforming diabetes care, allowing for more precise blood glucose management (Yu et al., 2024).

The following maps (Figs. 4.3 and 4.4) illustrate the age-adjusted distribution of diabetes in the study area. Readers interested in exploring the data sources can refer to the Spatial software and datasets section in the Appendix of Chapter One. Figure 4.3 is a choropleth map showing the prevalence of diabetes in each county as a percentage where darker shaded colors signify higher values while lighter shaded colors signify lower values. Fig. 4.4 is a hot spot analysis map where red denotes statistically significant clusters of high prevalence values (i.e., counties with high percentages adjacent to other counties with high percentages) while blue denotes

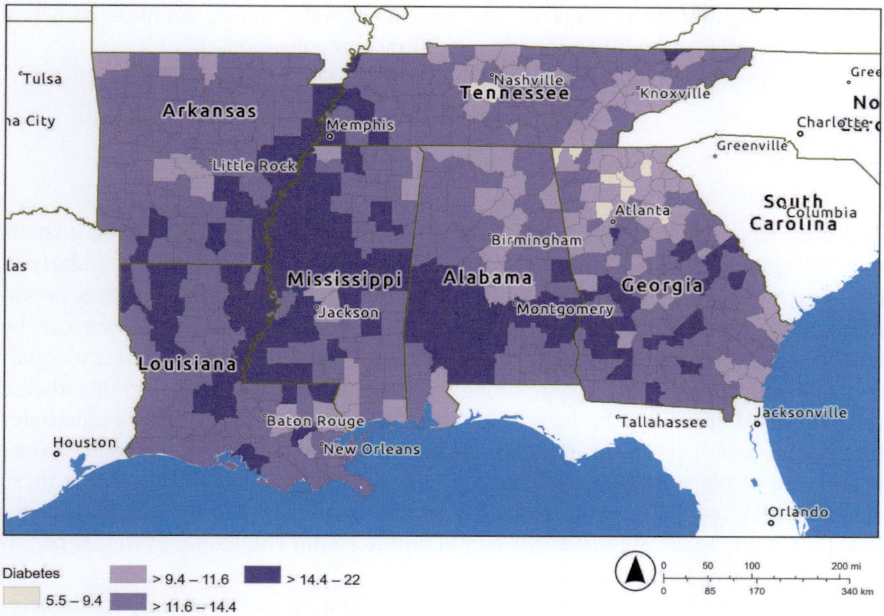

Fig. 4.3 Diabetes (percent prevalence). Source: ESRI, TomTom, Garmin, FAO, NOAA, USGS, EPA, USFWS (A. Sharma)

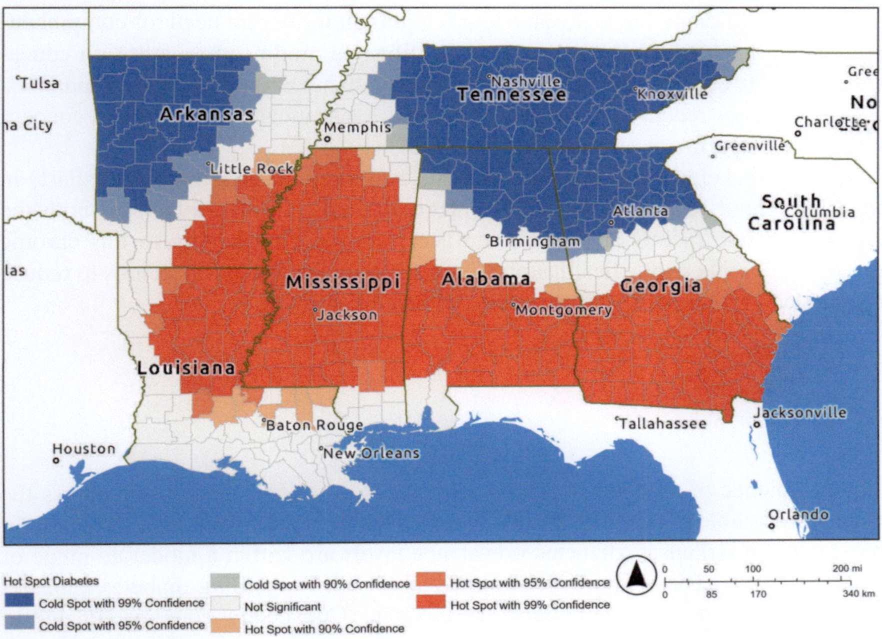

Fig. 4.4 Diabetes hot spots. Source: ESRI, TomTom, Garmin, FAO, NOAA, USGS, EPA, USFWS (A. Sharma)

statistically significant clusters of low prevalence values (i.e., counties with low percentages adjacent to other counties with low percentages).

Alabama

The prevalence of diabetes across Alabama reveals significant geographic disparities, as depicted in Fig. 4.3. Notably, the southern region of the state, particularly in counties adjacent to Montgomery, demonstrates a strikingly high diabetes prevalence with values ranging from 14.4% to 22%. This elevated prevalence can be attributed to a complex interplay of socio-economic factors, such as income inequality and educational attainment, along with disparities in access to healthcare resources. Additionally, the lack of health education in these regions exacerbates this public health issue. In contrast, Northern Alabama presents a slightly lower prevalence of diabetes with figures ranging between 9.4% and 11.6%. While these statistics may indicate a more favorable situation compared to the southern counties, they nonetheless signify a substantial public health challenge given this pervasive condition on a broader scale.

Major urban centers, including Birmingham and Montgomery, report diabetes prevalence levels that are comparatively higher, approximating within the range of 11.6% to 22%. This observation is particularly concerning given that urban areas typically offer greater access to healthcare services. The persistence of high diabetes levels in these metropolitan regions suggests that merely increasing healthcare access is insufficient. Such elevated levels highlight the urgent need for comprehensive public health initiatives that focus on lifestyle modification, nutrition education, and early detection to mitigate the impact of diabetes within these communities. As of the most recent data, diabetes remains a significant public health concern (Alabama Department of Public Health, 2024).

Overall, the elevated prevalence of diabetes throughout Alabama, particularly in the highlighted regions, calls for an integrated approach that addresses both the socio-economic factors and healthcare access issues contributing to this chronic disease. Effective public health strategies must consider these dimensions to reduce the burden of diabetes and improve health outcomes across the state.

Arkansas

The prevalence of diabetes in Arkansas displays a uniform distribution across the state with notable exceptions in eight counties located in the southeastern region. In most parts of Arkansas, diabetes prevalence levels are within a moderate range of 9.4% to 11.6%. These figures indicate a stable status regarding diabetes management and awareness in the majority of the state. Conversely, counties situated in proximity to the Mississippi Delta report higher diabetes prevalence levels, ranging from 14.4% to 22%. This discrepancy raises considerable public health concerns, particularly in areas such as Arkansas City and Helena, which exhibit elevated levels of diabetes.

Certain counties in southwestern Arkansas parallel this trend of elevated diabetes prevalence with rates consistently exceeding 14%. The comparative analysis of diabetes prevalence across these diverse regions underscores the necessity of acknowledging geographic disparities in health outcomes. These disparities can inform targeted public health interventions and resource allocations to address the pressing health challenges faced by affected communities. A comprehensive understanding of the geographic distribution of diabetes prevalence in Arkansas is essential for developing effective health policies and strategies aimed at improving overall health outcomes and reducing the burden of diabetes in the state.

Georgia

The prevalence of diabetes in Georgia exhibits significant variability compared to other states within the study area. In Northern Georgia, which encompasses urban centers such as Atlanta, the prevalence of diabetes is relatively low with estimates ranging from 5.5% to 11.6%. In contrast, Central Georgia reports higher prevalence levels with figures fluctuating between 9.4% and 14.4%. The southeastern region of the state displays a similar trend but still maintains lower rates compared to its southwestern counterpart, which presents elevated prevalence figures ranging from 14.4% to 22%. These data highlight a marked regional disparity in diabetes prevalence across Georgia. Northern Georgia, characterized by its urban infrastructure and healthcare resources, reflects a comparatively lower prevalence of diabetes. Northern Georgia has seen a significant increase in healthcare accessibility, which facilitates timely medical interventions and regular health monitoring for individuals at risk of developing diabetes. This enhanced access allows for early detection, better management of the condition, and a comprehensive approach to prevention. Conversely, the higher levels observed in the southern regions underscore a pressing public health concern that may be linked to several factors, including access to healthcare, lifestyle choices, socio-economic status, and prevalence of risk factors associated with diabetes. Community-based educational initiatives have proven instrumental in raising awareness about diabetes risk factors and the importance of maintaining a balanced diet. Additionally, initiatives that encourage physical activity and offer assistance with weight management have shown positive outcomes. Collaboration between local organizations, healthcare providers, and public health agencies has strengthened efforts to address diabetes on multiple fronts, ensuring that residents not only receive medical care but also engage in preventive measures.

Louisiana

The prevalence of diabetes in Louisiana is significantly elevated, particularly in the northwestern and northeastern regions, as well as the often-highlighted Mississippi Delta area. In these specific locations, the statistics are alarming with prevalence levels typically ranging from 14.4% to as high as 22%. This concentration of diabetes cases not only raises concerns for the affected communities but also points to a broader regional health crisis that extends into neighboring states, such as Arkansas and Mississippi. The geographical overlap of these high prevalence areas underscores the urgent need for collaborative health initiatives that can effectively address this public health challenge on a regional scale. Moreover, public health practitioners must recognize that these high diabetes levels are compounded by an increased occurrence of coronary heart disease, which further complicates the health landscape in these regions. Together, these conditions create a dual health burden that demands immediate attention and resources.

In contrast, the rest of Louisiana exhibits a moderate to high diabetes prevalence with levels between 11.6% and 14.4%. While these figures are lower than those in the most affected areas, they still indicate a significant health issue that cannot be overlooked. The need for comprehensive public health strategies, community education, and preventative measures is essential to address diabetes and its related complications throughout the state. By promoting collaboration beyond state borders and unifying healthcare initiatives, Louisiana can make considerable progress in enhancing health outcomes for its residents.

Mississippi

Mississippi, in conjunction with its neighboring states Arkansas and Louisiana, is notable for its alarmingly elevated levels of diabetes, making it a significant public health concern in the southeastern United States. In half of the counties throughout Mississippi, diabetes prevalence fluctuates between a staggering 14.4% and 22%. This pervasive issue highlights the urgent need for widespread health interventions and educational initiatives. What is especially concerning is the situation in the Mississippi Delta region, an area characterized by socio-economic challenges and limited access to healthcare resources. In this region, approximately 30 counties report diabetes prevalence rates near 20%. The concentration of diabetes in this area is not merely a statistic for it reflects broader systemic health disparities influenced by factors such as poverty, nutrition, and access to medical care.

The elevated levels of diabetes in Mississippi serve as a call for healthcare providers, community organizations, policymakers, and business leaders to implement comprehensive strategies that address the causes of this chronic illness. Efforts to improve health education, enhance access to nutritious foods, and promote physical activity are critical to reversing this troubling trend and improving the overall health outcomes for residents of Mississippi and the Delta region.

Tennessee

The prevalence of diabetes in Tennessee shows significant variation when compared to neighboring states in the region. In Western Tennessee, notably in urban areas such as Memphis, the prevalence is alarmingly high, ranging between 14.4% and 22%. This concentration of diabetes cases highlights a pressing public health concern that requires community-level intervention strategies. Moving to Central Tennessee, the statistics are moderate with diabetes prevalence from 9.4% to 11.6%. However, this figure begins to rise slightly in the areas east of Nashville, where levels increase between 11.6% and 14%. This gradient suggests that while Central Tennessee experiences lower diabetes levels than its western counterpart, it still faces challenges that warrant ongoing monitoring and healthcare initiatives. In contrast, Eastern Tennessee, which encompasses regions around Knoxville, reports a relatively lower prevalence of diabetes at 9%. This figure indicates a favorable situation compared to the more densely populated and urbanized western areas.

Overall, the variation in diabetes prevalence across Tennessee reflects not only geographical disparities but also suggests underlying socio-economic factors, healthcare accessibility, and lifestyle choices may contribute to these differences. Addressing these issues will be central for improving health outcomes and reducing the burden of diabetes across the state.

Hot Spot Analysis

Figure 4.4 presents the results of a comprehensive statistical analysis, illustrating distinct clusters of diabetes prevalence across the study area. These clusters are categorized with robust confidence levels set at 90%, 95%, and 99%. As previously noted, the shading on the map employs darker shades of red (or blue) to signify areas with high confidence and lighter shades of red (or blue) to represent lower confidence counterparts.

The analysis reveals a significant concentration of high-value clusters throughout the study region, particularly in counties where diabetes prevalence is approximately 22%. Key areas identified include southeast Arkansas, northeast Louisiana, the entirety of Mississippi, the southern region of Alabama, and both central and southern Georgia. These areas collectively highlight the alarming prevalence of diabetes, underscoring it as a pressing public health concern. Conversely, low-value clusters are observed in regions such as Northeast Arkansas, all of central and eastern Tennessee, the northern segment of Alabama, and the northern parts of Georgia. These clusters suggest lower diabetes prevalence, presenting a stark contrast to the high-value regions.

In summary, the hot spot analysis underscores that diabetes is not merely a localized issue but a widespread public health challenge affecting various communities across the study area. The data reveal alarming patterns of diabetes prevalence that vary by geography, race and ethnicity, health professional shortage areas, and rural and urban settings (refer to maps from previous chapters). This insight calls for targeted intervention strategies and resources to address the growing diabetes epidemic effectively.

Obesity

Obesity is increasingly recognized as a critical global public health issue. It is characterized by an excessive accumulation of body fat and is often assessed utilizing the Body Mass Index (BMI) with a BMI of 30 or greater serving as an indicator of obesity (WHO, 2024a, 2024b).[1] In January 2025, a group of 58 international scholars, designated as the Commission, proposed a redefinition of obesity that includes additional measures beyond the conventional BMI assessment (Endocrinology, 2025). The Commission recommended the application of BMI primarily as a screening instrument for population-level assessments rather than for individual-level risk evaluation. While BMI serves as a useful baseline measure, current research suggests that intramuscular adipose tissue (IMAT) may more accurately account for individual variations in fat distribution and could be a significant determinant in evaluating cardiovascular risk (Souza et al., 2025). IMAT, which is distributed within muscle fibers or near muscle groups, is commonly quantified using computed tomography (CT) or magnetic resonance imaging (MRI). When assessing IMAT, health practitioners should also consider its location and type.

The obesity epidemic poses considerable implications for both individual and public health systems, contributing to an increased incidence of chronic diseases and escalating healthcare costs. In the United States, the prevalence of obesity has risen dramatically over recent decades. The Centers for Disease Control and Prevention (CDC) reports that approximately 41.9% of American adults were classified as obese during the 2017–2020 period, a significant increase from 30.5% in 1999–2000 (CDC, 2024a). Contributing factors include sedentary lifestyles, poor dietary choices, and various environmental influences. In the Southeastern United States, the issue of obesity is particularly pronounced with alarming prevalence levels and associated health complications. This region faces distinct challenges that exacerbate its high obesity rates. According to the CDC, states in the Southeast report adult obesity rates exceeding 35% with Mississippi, Arkansas, Louisiana, and Alabama consistently ranking among the highest and #2–5 overall for the nation, respectively (see Table 4.4; CDC, 2024b).

---

[1] BMI is measured as weight (kg)height2 (meters)

Table 4.4 Obesity prevalence in the study area (2023)

State	Prevalence percentage
Mississippi	40.1
Arkansas	40.0
Louisiana	39.9
Alabama	39.2
Tennessee	37.6
Georgia	35.0

Source: Adult obesity prevalence maps based on Behavioral Risk Factor Surveillance System, BRFSS 2023 (https://www.cdc.gov/obesity/data-and-statistics/adult-obesity-prevalence-maps.html)

Cultural factors play a role in shaping dietary behaviors across the Southeast. The region's traditional cuisine is characterized by the prevalence of high-calorie and high-fat foods, including buttermilk biscuits, eggs, fried dishes, and rich sauces. These dietary patterns are associated with increasing rates of obesity (Couch et al., 2021). Additionally, social gatherings in this region frequently center around food, leading to increased portion sizes and greater calorie intake. Societal norms and environmental factors further influence these eating behaviors (Couch et al., 2021).

Socio-economic factors complicate the obesity landscape in the Southeast. Low-income populations often face limited access to nutritious food and safe environments for physical activity. Studies indicate that food deserts—areas with restricted access to affordable and healthy food—are prevalent in many Southeastern and rural communities, restricting residents' ability to maintain a healthy diet and contributing to elevated obesity rates (Champagne et al., 2007; Larson et al., 2009; Oh et al., 2024; Wilcox et al., 2020). Additionally, many communities in the Southeast lack recreational facilities, safe walking paths, and parks, leading to reduced levels of physical activity (Brown et al., 2023). The interplay of unhealthy dietary patterns and sedentary lifestyles creates a critical environment for rising obesity rates.

Obesity is linked to numerous health issues, including Type 2 diabetes, hypertension, cardiovascular diseases, and certain cancers (Vucenik et al., 2024). The correlation between obesity and mental health is also gaining attention, as individuals with obesity may experience heightened rates of depression and anxiety (Luppino et al., 2010). The economic implications are substantial with the CDC estimating that obesity-related healthcare costs amount to approximately $173 billion annually, underscoring the urgent need for effective intervention strategies (CDC, 2024A).

Addressing obesity necessitates a comprehensive approach that encompasses individual-level interventions, community-based programs, and policy changes. Key strategies include enhancing access to healthy foods, promoting physical activity, and implementing educational initiatives focused on nutrition (Jilcott et al.,

2011). For instance, community gardening and farmers' markets have demonstrated effectiveness in increasing fruit and vegetable intake among low-income populations (Freedman et al., 2013; Schupp, 2016). Targeted policy measures, such as taxes on sugary beverages and regulations on food advertising, can facilitate the creation of healthier environments (Brownell & Frieden, 2009). Additionally, policies that promote active living spaces—such as the development of parks and walking trails—are essential for mitigating the obesity epidemic (Verde et al., 2024).

The subsequent maps (Figs. 4.5 and 4.6) detail the age-adjusted distribution of obesity within the study area. Readers interested in exploring the data sources can refer to the Spatial software and datasets section in the Appendix of Chapter One. Figure 4.5 provides a choropleth visualization of obesity prevalence by county, where darker shades represent higher prevalence levels and lighter shades indicate lower levels. In contrast, Fig. 4.6 utilizes hot spot analysis to identify statistically significant clusters: red areas denote regions with high prevalence adjacent to other high prevalence counties while blue areas indicate clusters of low prevalence in proximity to other low prevalence counties.

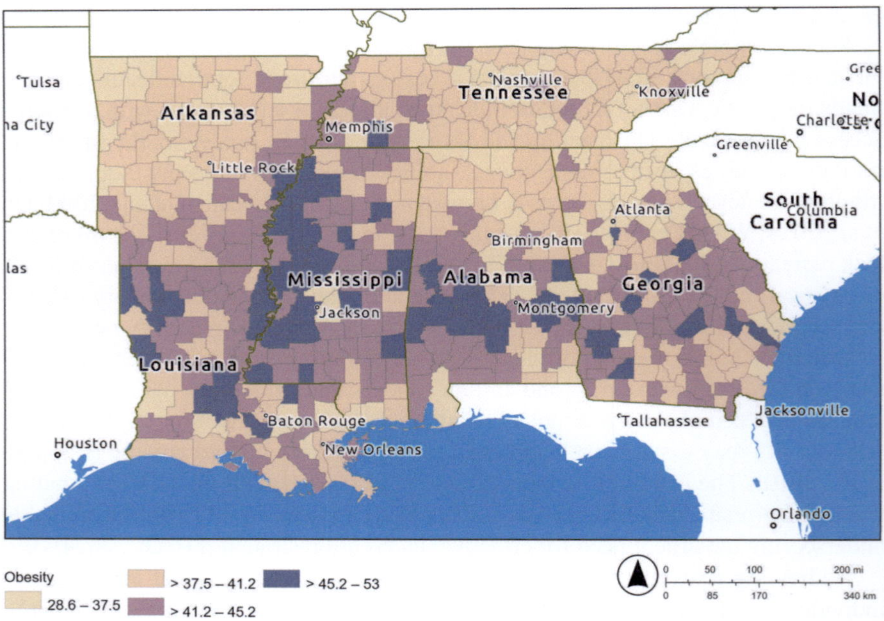

Fig. 4.5 Obesity (percent prevalence). Source: ESRI, TomTom, Garmin, FAO, NOAA, USGS, EPA, USFWS (A. Sharma)

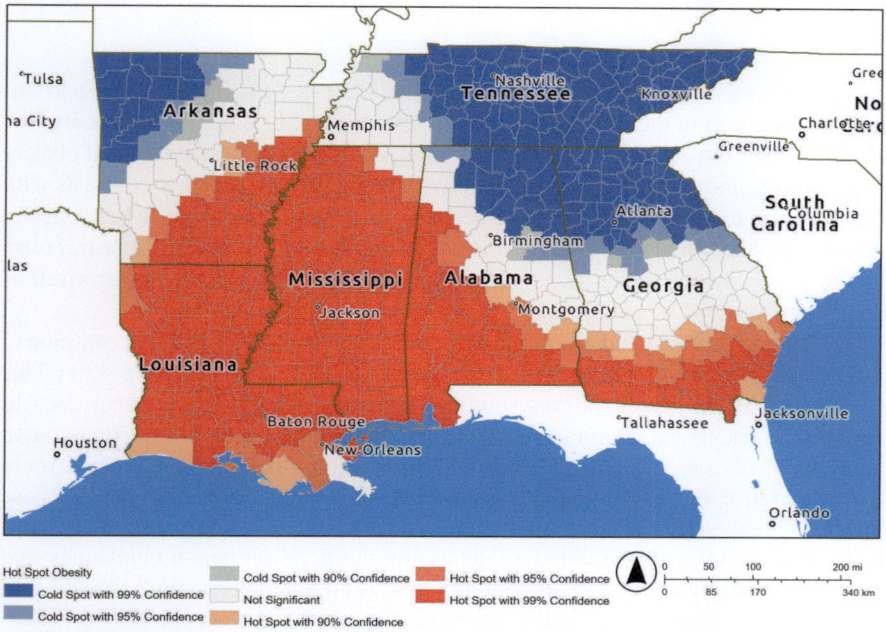

Fig. 4.6 Obesity hot spots. Source: ESRI, TomTom, Garmin, FAO, NOAA, USGS, EPA, USFWS (A. Sharma)

Alabama

The prevalence of obesity in Alabama exhibits significant geographic variation, as illustrated in Fig. 4.5. The southern region, particularly in counties bordering Mississippi, shows a notably high obesity prevalence with levels ranging from 40.9% to 44.7%. Higher levels of poverty and the insufficient availability of preventive healthcare services and health education in these areas exacerbate this public health challenge. Conversely, Northern Alabama reflects a slightly lower obesity prevalence with levels below 37%. Although this situation appears more favorable compared to the southern counties, it still represents a significant public health concern. Overall, the elevated obesity prevalence throughout Alabama, particularly in the southern region, necessitates a comprehensive approach that addresses both socio-economic determinants and healthcare access issues associated with this chronic disease. Effective public health strategies must take these factors into account to alleviate the burden of obesity and improve health outcomes statewide.

Arkansas

The prevalence of obesity in Arkansas reveals marked geographic variation, particularly pronounced in the southeastern region of the state. In other areas, obesity levels range between 30% and 37.2%, suggesting a stable landscape in terms of obesity management and public awareness initiatives. However, this stability contrasts with the alarming figures reported in counties near the Mississippi Delta, where obesity prevalence peaks at levels ranging from 40.9% to over 44.7%. Such disparities raise urgent public health concerns, especially in communities which are categorized as health professional shortage areas.

In addition to the Delta region, several counties in southern Arkansas continue to reflect this trend of elevated obesity prevalence with levels exceeding 40%. This consistent pattern across various southern regions highlights the critical need to address the underlying factors contributing to these health disparities. Comparative analyses of obesity prevalence across the northern and southern areas underscore the importance of recognizing geographic disparities in health outcomes. These insights can guide the development of targeted public health interventions and strategic resource allocations aimed at addressing the pressing health challenges that disproportionately affect certain communities. A comprehensive understanding of the geographic distribution of obesity prevalence in Arkansas is necessary for crafting effective health policies and strategies.

Georgia

The obesity epidemic in Georgia presents a notable geographic pattern with disparities between the northern and southern regions of the state. In Northern Georgia, which includes metropolitan hubs such as Atlanta, obesity levels are comparatively lower, ranging from 30% to 37.2%. This region benefits from greater urban infrastructure and enhanced healthcare resources that contribute to a more moderate prevalence of obesity. The increase in healthcare accessibility has been central in facilitating timely medical interventions and regular health monitoring for at-risk individuals by enabling early detection and better management of obesity. With an emphasis on holistic approaches to prevention, the community in Northern Georgia tends to have better tools and resources for promoting healthy lifestyles.

In Central Georgia, the obesity prevalence escalates with estimates ranging between 37.2% and 40.9%. This increase reflects a trend that highlights a transition rather than a sharp divide, indicating that although Central Georgia may not be as affected as the southern part, it still contends with rising obesity levels that require attention. In stark contrast, the southern region of the state exhibits a troubling trend with some counties reporting obesity levels exceeding 44.7%. These elevated levels signify a pressing public health crisis that is influenced by various interrelated factors. Access to healthcare may be limited in these areas, combined with lifestyle choices that may not prioritize nutrition and physical activity. Socio-economic challenges further exacerbate these issues, creating an environment where risk factors associated with obesity are more prevalent and difficult to counteract.

Recognizing the urgent need to address this disparity, community-based educational initiatives have emerged as effective tools for raising awareness about the risk factors of obesity, particularly in rural areas. These programs highlight the important role of a balanced diet that includes a variety of nutritious foods. Collaboration between local healthcare providers, community organizations, and public health agencies has been essential in launching comprehensive efforts to combat obesity. This multifaceted approach not only ensures that residents receive necessary medical care but also empowers them to adopt preventive measures that can lead to long-term health improvements.

Louisiana

The prevalence of obesity in Louisiana is alarmingly high with many regions reporting levels between 40.9% and 44.7%. This concentration of obesity cases is particularly concerning for rural communities designated as health professional shortage areas as obesity is a precursor to numerous health complications, including the previously mentioned cardiovascular diseases and diabetes. The burden of these health issues places immense strain on local healthcare systems, leading to increased medical costs and reduced availability of resources. That is, high obesity levels can create a cycle of poor health outcomes, contributing to higher rates of hospitalization and poor chronic disease management. Additionally, the implications of obesity influence mental well-being and the overall quality of life for residents.

In Northern Louisiana, the situation is even more severe with obesity levels exceeding 44.7%. This worrisome estimate not only intensifies public health issues but also emphasizes the urgent need for targeted interventions and community support initiatives. Addressing the obesity epidemic in this region requires comprehensive strategies that promote collaboration among various stakeholders. Healthcare practitioners, business leaders, community organizations, and policymakers must work collectively to create effective solutions. Potential strategies might include enhancing access to nutritious food options, promoting physical activity through community programs during the many statewide celebrations and festivals, implementing educational campaigns focused on healthy lifestyles, and encouraging supportive policies that promote wellness. Furthermore, community engagement is vital; residents must be active participants in the formulation of these initiatives to ensure they are culturally relevant and effectively meet the needs of the population. By taking a multifaceted approach that involves collaboration, education, and community empowerment, Louisiana can manage its obesity crisis in a practical and cost-effective manner.

Mississippi

Mississippi is recognized for its distressingly elevated levels of obesity, which have become a public health crisis. In many areas of the state, the obesity prevalence exceeds 40%, underscoring a troubling pattern that demands immediate attention. While some counties in the vicinity of Jackson present slightly more manageable levels—ranging from 30% to 37.2%—the overall landscape remains alarming.

The widespread obesity levels can be attributed to several factors, including the socio-economic challenges prevalent in several rural counties. In these areas, access to nutritious food, walkable neighborhoods, and educational opportunities are often limited, thereby contributing to poor dietary habits and sedentary lifestyles. As outlined in Chapter One, the classification of these areas as rural adds complexity to the issue. Individuals living in rural Mississippi may contend with food deserts. These are regions where fresh fruits and vegetables are scarce and convenient access to healthy food options is often nonexistent. Moreover, transportation challenges further complicate matters, making it difficult for residents to reach supermarkets or healthcare facilities. This isolation can perpetuate a cycle of unhealthy eating habits and prevent individuals from seeking medical advice or preventative care. In addition, cultural attitudes toward food and lifestyle choices can also play a role in shaping health outcomes. Traditional Southern cuisines, which are often rich in calories, fats, and sugars, often dominate dining tables, thereby complicating efforts to embrace healthier eating habits. Combined with a tendency towards less physically active lifestyles due to weather and geographical factors, these aspects contribute to elevated obesity levels.

Addressing Mississippi's obesity crisis will require comprehensive and long-term approaches that involve the collaboration of various stakeholders, including local governments, health organizations, educational institutions, and community groups. Effective strategies could include launching farmers' markets, establishing community gardens, and forming partnerships with local retailers to make healthy food options more accessible and affordable. Furthermore, enhancing transportation options is essential as it allows residents to access grocery stores and healthcare facilities more conveniently.

Tennessee

The prevalence of obesity in Tennessee presents an intriguing landscape, exhibiting lower levels than those found in several neighboring states. Variations in obesity levels can be observed when one examines different areas of the state. In Western Tennessee, especially in urban centers such as Memphis, obesity levels are elevated, ranging from 30% to a notable 40.9%. This higher prevalence in urban settings may be influenced by factors such as lifestyle choices, food availability, and economic conditions prevalent in more populated areas. As we move towards Central

Tennessee, the distribution becomes more uniform with obesity levels ranging between 30% and slightly over 37.2%. This moderate level reflects a blend of urban and rural influences, suggesting that both environments play a role in shaping health outcomes. Within this region, Nashville serves as an exception where obesity levels approach 37.2%, which is lower than those in the other major city of Memphis.

In Eastern Tennessee, particularly around Knoxville, there is a noticeable decline in obesity prevalence with estimates below 37.2%. The relatively lower levels in this area indicate a more favorable health scenario compared to the denser urban populations found in the western part of the state. This disparity suggests that factors such as rural lifestyle, community health programs, and better access to recreational spaces may contribute to this more health-conscious environment. Overall, the trend of lower obesity prevalence throughout Tennessee not only highlights geographical disparities but also points to a complex interplay of socio-economic factors, healthcare access, and lifestyle choices.

Hot Spot analysis

Figure 4.6 displays the outcomes of an extensive statistical analysis that delineates distinct clusters of obesity prevalence within the study area. These clusters are classified into high and low-value categories and are supported by robust confidence levels of 90%, 95%, and 99%. The map uses a color gradient with darker shades of red (or blue) indicating areas of higher confidence and lighter shades of red (or blue) representing regions with lower confidence levels.

The analysis reveals a pronounced concentration of high-value clusters across the study region, particularly in counties where the obesity prevalence is estimated at approximately 44.7%. Notable areas include southeast Arkansas, northeast Louisiana, the majority of Mississippi—excluding its northeastern and southeastern sections—and the southern region of Alabama. Collectively, these regions underscore the critical prevalence of obesity, positioning it as a significant public health challenge. In contrast, low-value clusters are identified in areas such as Northern Georgia and Central and Eastern Tennessee. These regions indicate a lower prevalence of obesity, starkly contrasting with their high-value counterparts.

In conclusion, the geographical analysis reveals that obesity is not merely a localized concern but constitutes a widespread public health crisis impacting diverse communities in Alabama and Mississippi, as well as the previously mentioned greater Mississippi Delta region. This compelling insight highlights the urgent need for intervention strategies and the allocation of resources to effectively combat the escalating obesity epidemic.

Concluding Remarks

Chronic diseases pose a public health challenge in the Southeastern United States, a region characterized by disproportionately high rates of several leading causes of mortality. Elevated obesity rates in many counties within this area further exacerbate the public health crisis, as obesity is a contributing factor to conditions such as coronary heart disease (CHD) and diabetes. Recent analyses employing choropleth and hot spot mapping techniques have revealed concerning levels of CHD, diabetes, and obesity, particularly in the greater Mississippi Delta region. These visualizations highlight critical health disparities experienced by this community and underscore the pressing need for effective and innovative public health strategies.

To improve health outcomes and reduce mortality rates among this vulnerable population, addressing the interrelated nature of these chronic conditions is essential. Mitigating these issues necessitates a multifaceted approach, incorporating policy reforms, community education, and improved access to healthcare resources to combat the rising prevalence of obesity and its associated health complications (Ivassenko et al., 2022). Below, I present key initiatives undertaken to enhance population health and well-being with the hope that healthcare practitioners and policymakers will continue to advocate for and expand these crucial programs.

Alabama: Acknowledging the need to address rural health challenges, the Department of Public Health has designated the third Thursday of November as National Rural Day. In November 2022, Governor Ivey formally proclaimed this day as Rural Health Day in Alabama. This observance highlights the distinctive health challenges faced by rural communities and honors the collaborative efforts of individuals and organizations dedicated to enhancing healthcare access and quality (see https://www.alabamapublichealth.gov/ruralhealth/nrhd.html).

Arkansas: In 2015, the Arkansas Center for Health Improvement (ACHI) launched the Healthy Active Arkansas initiative and aimed at reducing obesity and enhancing physical activity levels among residents through the implementation of public policies and community engagement. Endorsed by the Governor's office, this program emphasizes the critical role of collaboration among diverse stakeholders to develop effective action plans that foster healthy lifestyles throughout the state (see https://achi.net/our-initiatives/healthy-active-arkansas/).

Georgia: Recognizing the importance of cultivating healthy behaviors during early childhood to enhance well-being, Georgia initiated the SHAPE program in 2012. This comprehensive statewide program aimed to increase physical activity and improve nutrition within schools with a long-term objective of mitigating childhood obesity through the implementation of wellness policies and the provision of resources designed to foster healthier educational environments (for further details, please visit https://www.georgiashape.org/). Concurrently, the Department of Public Health has introduced the Healthy Communities Initiative (HCI), which allocates funding to local communities to create supportive environmental and social systems that encourage increased nutrition and physical activity (for more information, see https://dph.georgia.gov/chronic-disease-prevention/healthy-communities-initiative).

Louisiana: The state takes immense pride in its rich cultural heritage and lifestyle while also acknowledging the impact of dietary choices on chronic health conditions, such as coronary heart disease, diabetes, and obesity. In response, the Department of Public Health established Well-Ahead Louisiana, a dedicated unit focused on chronic disease prevention and healthcare access. Well-Ahead collaborates with nearly 50 stakeholders statewide to promote exercise, housing improvements, nutritional health, tobacco cessation, and overall well-being (see https://wellaheadla.com/).

Mississippi: Consistent with initiatives in other states within our study area, Mississippi has established innovative programs focused on the objective of reducing obesity as a means to manage chronic diseases effectively. One notable initiative is UProot, a collaborative effort involving the Department of Public Health and nearly 90 businesses, community organizations, nonprofits, and state agencies (see https://uprootms.org/about/). UProot employs a comprehensive strategy that addresses food insecurity while promoting healthier lifestyles among the population.

Tennessee: The Department of Health is dedicated to enhancing population health and oversees a diverse portfolio of over fifty program areas. This considerable investment in public health is a key factor in the state's comparatively favorable health outcomes in the study area (see https://www.tn.gov/health.html). Among the notable programs are the Commodity Supplemental Food Program, Diabetes Management, Family Health and Wellness initiatives, the Farmer's Market Nutrition Program, and efforts focused on Heart Disease and Stroke prevention, including the Office of Primary Prevention. Furthermore, the Department of Health maintains comprehensive health data, providing health practitioners and policymakers with essential information to support effective planning and decision-making.

This chapter is intended to provide readers with a summary of the most prevalent health issues within the study area. I encourage individuals contemplating a career in health or those already engaged in the field to examine the department of public health website to gain a comprehensive understanding of the health objectives specific to each state. As highlighted in Chapters Two and Three, understanding the cultural and historical contexts of each region equips healthcare practitioners and providers to deliver services that respect the individual and the community. While this chapter has provided a comprehensive examination of the current health issues, the subsequent chapter will shift its focus to exploring the emerging health concerns that are intricately associated with changes in adverse weather conditions. This focus is particularly relevant considering the unique geographical characteristics of the region, which may increase health system vulnerability resulting from natural hazard risk. This forthcoming chapter aims to provide a detailed analysis of these complex interactions between natural disasters, changing atmospheric conditions, and health outcomes. By examining these relationships, it seeks to contribute to the development of informed preparedness and response frameworks that are designed to address the specific vulnerabilities related to health systems.

References

Alabama Department of Public Health. (2024). *Diabetes*. Alabama Department of Public Health. https://www.alabamapublichealth.gov/diabetes/index.html

American Diabetes Association. (2020). Standards of medical care in diabetes—2020. *Diabetes Care, 43*(Supplement 1), S1–S232. https://doi.org/10.2337/dc20-Sint

Arbab-Zadeh, A., Nakano, M., Virmani, R., & Fuster, V. (2012). Acute coronary events. *Circulation, 125*(9), 1147–1156.

Brown, N. I., Stewart, L., Rogers, L. Q., Powell, M. A., Hardy, C. M., Baskin, M. L., & Pekmezi, D. (2023). Assessing the built environment, programs, and policies that support physical activity opportunities in the rural Deep South. *Preventive Medicine Reports, 33*, 102223.

Brownell, K. D., & Frieden, T. R. (2009). Ounces of prevention—the public policy case for taxes on sugared beverages. *New England Journal of Medicine, 360*(18), 1805–1808.

Centers for Disease Control and Prevention (CDC). (2024a). *Adult obesity facts*. CDC. https://www.cdc.gov/obesity/php/data-research/adult-obesity-facts.html

Centers for Disease Control and Prevention (CDC). (2024b). *Adult obesity prevalence map*. CDC. https://www.cdc.gov/obesity/php/data-research/adult-obesity-prevalence-maps.html

Champagne, C. M., Casey, P. H., Connell, C. L., Stuff, J. E., Gossett, J. M., Harsha, D. W., & Lower Mississippi Delta Nutrition Intervention Research Initiative. (2007). Poverty and food intake in rural America: diet quality is lower in food insecure adults in the Mississippi Delta. *Journal of the American Dietetic Association, 107*(11), 1886–1894.

Couch, C. A., Gray, M. S., Shikany, J. M., Howard, V. J., Howard, G., Long, D. L., & Judd, S. E. (2021). Correlates of a southern diet pattern in a national cohort study of blacks and whites: the reasons for geographic and racial differences in stroke (REGARDS) study. *British Journal of Nutrition, 126*(12), 1904–1910.

Davis, D., Kurz, E., Hooper, M. E., Atchan, M., Spiller, S., Blackburn, J., & Knight-Agarwal, C. (2024). The holistic maternity care needs of women with Gestational Diabetes Mellitus: A systematic review with thematic synthesis. *Women and Birth, 37*(1), 166–176.

Eizirik, D. L., Szymczak, F., & Mallone, R. (2023). Why does the immune system destroy pancreatic β-cells but not α-cells in type 1 diabetes? *Nature Reviews Endocrinology, 19*(7), 425–434.

ElSayed, N. A., Aleppo, G., Aroda, V. R., Bannuru, R. R., Brown, F. M., Bruemmer, D., & Gabbay, R. A. (2024). 10. Cardiovascular disease and risk management: standards of care in diabetes—2023. *Diabetes Care, 47*, S158.

Endocrinology, T. L. D. (2025). Redefining obesity: advancing care for better lives. *The Lancet Diabetes & Endocrinology, 13*, S2213–S8587.

Flora, G. D., & Nayak, M. K. (2019). A brief review of cardiovascular diseases, associated risk factors and current treatment regimes. *Current Pharmaceutical Design, 25*(38), 4063–4084.

Freedman, D. A., Choi, S. K., Hurley, T., Anadu, E., & Hébert, J. R. (2013). A farmers' market at a federally qualified health center improves fruit and vegetable intake among low-income diabetics. *Preventive Medicine, 56*(5), 288–292.

Garber, A. J. (2015). AACE/ACE comprehensive diabetes management algorithm 2015. *Endocrine Practice, 21*(4), 438–447.

Ivassenko, V., Macnab, A. J., Di Emidio, D., Hollederer, A., Iellamo, E. A., Cabuslay, J. I. B., & Sharma, A. (2022). Engaging with people and populations in health promotion research: A snapshot on participatory processes. In *Global handbook of health promotion research, Vol. 1: Mapping health promotion research* (pp. 93–112). Springer International Publishing.

Jilcott, S. B., Keyserling, T., Crawford, T., McGuirt, J. T., & Ammerman, A. S. (2011). Examining associations among obesity and per capita farmers' markets, grocery stores/supermarkets, and supercenters in US counties. *Journal of the American Dietetic Association, 111*(4), 567–572.

Kirkpatrick, C. F., Greaves, K. A., & Foster, E. (2024). Lifestyle interventions and atherosclerotic cardiovascular disease outcomes. In *Cardiovascular outcomes research: A clinician's guide to cardiovascular epidemiology and clinical outcomes trials* (pp. 167–200). Springer International Publishing.

Krittanawong, C., Khawaja, M., Virk, H. U. H., Escobar, J., Khalid, U., Birnbaum, Y., & Virani, S. S. (2024). Strategies for chronic coronary disease: A brief guide for clinicians. *NPJ Cardiovascular Health, 1*(1), 6.

Lamm, R. D., & Sharma, A. (2013). *Brave new world of healthcare revisited: What every American needs to know about our healthcare crisis*. Fulcrum Publishing.

Larson, N. I., Story, M. T., & Nelson, M. C. (2009). Neighborhood environments: disparities in access to healthy foods in the US. *American Journal of Preventive Medicine, 36*(1), 74–81.

Libby, P., Ridker, P. M., & Hansson, G. K. (2009). Inflammation in atherosclerosis. *Journal of the American College of Cardiology, 54*(23), 2129–2138.

Llerena-Velastegui, J., Zumbana-Podaneva, K., Velastegui-Zurita, S., Mejia-Mora, M., Perez-Tomassetti, J., Cabrera-Cruz, A., & Sanahuja-Montiel, C. (2024). Comparative efficacy of percutaneous coronary intervention versus coronary artery bypass grafting in the treatment of ischemic heart disease: A systematic review and meta-analysis of recent randomized controlled trials. *Cardiology Research, 15*(3), 153–168.

Lu, W. L., Yuan, J. H., Liu, Z. Y., Su, Z. H., Shen, Y. C., Li, S. J., & Zhang, H. (2024). Worldwide trends in mortality for hypertensive heart disease from 1990 to 2019 with projection to 2034: data from the Global Burden of Disease 2019 study. *European Journal of Preventive Cardiology, 31*(1), 23–37.

Luppino, F. S., de Wit, L. M., Bouvy, P. F., Stijnen, T., Cuijpers, P., Penninx, B. W., & Zitman, F. G. (2010). Overweight, obesity, and depression: a systematic review and meta-analysis of longitudinal studies. *Archives of General Psychiatry, 67*(3), 220–229.

Marrero, D. G., Ard, J., Delamater, A. M., Peragallo-Dittko, V., Mayer-Davis, E. J., Nwankwo, R., & Fisher, E. B. (2013). Twenty-first century behavioral medicine: a context for empowering clinicians and patients with diabetes: a consensus report. *Diabetes Care, 36*(2), 463–470. https://doi.org/10.2337/dc12-2305

Members, W. C., Virani, S. S., Newby, L. K., Arnold, S. V., Bittner, V., Brewer, L. C., & Williams, M. S. (2023). 2023 AHA/ACC/ACCP/ASPC/NLA/PCNA guideline for the management of patients with chronic coronary disease: A report of the American Heart Association/American College of Cardiology Joint Committee on Clinical Practice Guidelines. *Journal of the American College of Cardiology, 82*(9), 833–955.

NCD Risk Factor Collaboration. (2016). Worldwide trends in diabetes since 1980: A pooled analysis of 751 population-based studies with 4.4 million participants. *The Lancet, 387*(10027), 1513–1530.

Oh, J. I., Lee, K. J., & Hipp, A. (2024). Food deserts exposure, density of fast-food restaurants, and park access: Exploring the association of food and recreation environments with obesity and diabetes using global and local regression models. *PLoS One, 19*(4), e0301121.

Omran, A. R. (1971). The epidemiologic transition: A theory of the epidemiology of population change. *The Milbank Memorial Fund Quarterly, 49*(4), 509–538.

Omran, A. R. (2001). The epidemiologic transition: A theory of the epidemiology of population change. *Bulletin of the World Health Organization, 79*(2), 161–170.

Parker, E. D., Lin, J., Mahoney, T., Ume, N., Yang, G., Gabbay, R. A., & Bannuru, R. R. (2024). Economic costs of diabetes in the US in 2022. *Diabetes Care, 47*(1), 26–43.

Roden, M., Petersen, K. F., & Shulman, G. I. (2024). Insulin resistance in type 2 diabetes. In *Textbook of diabetes* (pp. 238–249).

Saydah, S., Bullard, K. M., Cheng, Y., Ali, M. K., Gregg, E. W., Geiss, L., & Imperatore, G. (2014). Trends in cardiovascular disease risk factors by obesity level in adults in the United States, NHANES 1999-2010. *Obesity, 22*(8), 1888–1895.

Schupp, J. L. (2016). Just where does local food live? Assessing farmers' markets in the United States. *Agriculture and Human Values, 33*, 827–841.

Souza, A. C. D. A., Troschel, A. S., Marquardt, J. P., Hadžić, I., Foldyna, B., Moura, F. A., Hainer, J., Divakaran, S., Blankstein, R., Dorbala, S., Di Carli, M. F., Aerts, H. J., Lu, M. T., Fintelmann, F. J., & Taqueti, V. R. (2025). Skeletal muscle adiposity, coronary microvascular dysfunction, and adverse cardiovascular outcomes. *European Heart Journal, 46*, 1112.

Thygesen, K., Alpert, J. S., Jaffe, A. S., Chaitman, B. R., Bax, J. J., Morrow, D. A., & White, H. D. (2019). Fourth universal definition of myocardial infarction (2018). *European Heart Journal, 40*(3), 237–269.

Verde, L., Barrea, L., Bowman-Busato, J., Yumuk, V. D., Colao, A., & Muscogiuri, G. (2024). Obesogenic environments as major determinants of a disease: It is time to re-shape our cities. *Diabetes/Metabolism Research and Reviews, 40*(1), e3748.

Vucenik, I., Jones, L. P., & McLenithan, J. C. (2024). Linking obesity, metabolism, and cancer. In *Metabolic syndrome: A comprehensive textbook* (pp. 603–620). Springer International Publishing.

Wilcox, S., Sharpe, P. A., Liese, A. D., Dunn, C. G., & Hutto, B. (2020). Socioeconomic factors associated with diet quality and meeting dietary guidelines in disadvantaged neighborhoods in the Southeast United States. *Ethnicity & Health, 25*(8), 1115–1131.

World Health Organization (WHO). (2024a). *Obesity and overweight*. WHO. https://www.who.int/news-room/fact-sheets/detail/obesity-and-overweight

World Health Organization (WHO). (2024b). *Diabetes facts*. WHO. https://www.who.int/news-room/fact-sheets/detail/diabetes

World Health Organization (WHO). (n.d.). *Diabetes*. WHO. https://www.who.int/health-topics/diabetes#tab=tab_1

Yu, T. S., Song, S., Yea, J., & Jang, K. I. (2024). Diabetes management in transition: Market insights and technological advancements in CGM and insulin delivery. *Advanced Sensor Research, 3*, 2400048.

Chapter 5
Emerging Health Issues and Natural Disasters

Abstract Natural hazards, including floods, hurricanes, severe storms, and tornadoes, disrupt healthcare systems and compromise the accessibility and continuity of health services. As the impacts of adverse weather events become increasingly noticeable in the Southeastern United States and worldwide, health professionals, emergency management teams, and policymakers must engage in collaborative efforts throughout the phases of disaster preparedness, response, and recovery. This chapter examines the challenges presented by natural disasters with a particular emphasis on their effects within rural communities and the emergence of associated health concerns. Furthermore, the chapter introduces the lifelines framework, the Social Vulnerability Index (SVI), and other resources and tools to support an evaluative approach. I also provide an application in the context of kidney dialysis treatment to emphasize the intersection of social vulnerability and healthcare delivery during disaster events.

Keywords Natural disasters · Natural hazards · Emergency management · Disaster events · Health systems · Health outcomes · Resilience · Social Vulnerability

---

This chapter, authored by **Amanda Sharma**, addresses emerging health challenges confronting rural communities due to adverse weather events and natural disasters. The perspective presented is informed by a diverse background spanning the U.S. EPA, FEMA, and local government, with experience in management and planning roles focused on hazard mitigation, disaster recovery, and environmental health outreach. As founder of Place Mosaic LLC, Amanda Sharma supports government agencies and other organizations through risk communication, planning, and systems integration. Credentials include an MBA in Executive Leadership and a Master of Resource Law Studies from the University of Denver, along with certifications in leadership coaching, team coaching, and change management. Field experience includes deployments to multiple national disasters and working knowledge of the National Incident Management System (NIMS) and the Incident Command System (ICS).

A. Sharma, *Current and Emerging Rural Health Issues in the Southeast United States*, Global Perspectives on Health Geography,
https://doi.org/10.1007/978-3-032-07543-7_5

Introduction

This chapter explores the emerging health challenges associated with adverse weather events with a specific focus on natural disasters and community health systems. Through a comprehensive review of the relevant literature and empirical research, this chapter aims to identify strategies for enhancing community preparedness and response to these evolving health challenges. It highlights the predominant adverse weather events and natural disasters that have affected Alabama, Arkansas, Georgia, Louisiana, Mississippi, and Tennessee over the past twenty years. For purposes of this chapter, adverse weather events refer to meteorological conditions that can have detrimental impacts on the environment and human health due to their duration, frequency, and intensity. These events are characterized as adverse due to their capacity to disrupt community functioning, damage infrastructure, and compromise public safety. Changing atmospheric conditions refers to "a change in the state of the climate that can be identified (e.g., by using statistical tests) by changes in the mean and/or the variability of its properties and that persists for an extended period, typically decades or longer" (Lee et al., 2023). A natural disaster can be defined as an adverse weather event characterized by sudden onset and limited duration, which leads to a disaster declaration issued by a state or tribal government. Such an event must be formally recognized by the President of the United States as eligible for federal assistance under the Robert T. Stafford Disaster Relief and Emergency Assistance Act (FEMA, n.d.-a). The primary types of adverse weather events resulting in federally declared disasters in these states include floods, hurricanes, severe storms, and tornadoes.

Storms, Hurricanes, Floods, and Tornadoes

Before discussing the adverse impact of weather incidents on health systems, readers will benefit by first understanding how these events are defined. Severe storms can manifest as thunderstorms accompanied by rain and lightning, which can lead to flooding and fires. These storms can be amplified by factors such as high winds, hail, and tornado activity. The National Oceanic and Atmospheric Administration (NOAA) classifies a storm as "severe" when it generates wind gusts of at least 58 miles per hour (mph) and/or produces hail one inch in diameter (comparable to a U.S. quarter coin) or larger, and/or results in the formation of a tornado (NOAA, n.d.).

Hurricanes, a distinct type of tropical cyclone that develops over tropical or subtropical waters, are classified by the National Hurricane Center (NOAA, 2024). The severity of hurricanes is assessed using the Saffir-Simpson Hurricane Wind Scale, which ranges from Category 1 to 5 based on sustained wind speeds. Category 1 begins at winds of 75 mph while Category 5 encompasses wind speeds exceeding 157 mph. The primary threats from hurricanes are high winds and storm surges, both of which can cause significant damage. In our study area, three states—Louisiana, Georgia, and Mississippi—adjoin the Gulf Coast, yet hurricane impacts are felt beyond coastal regions. For instance, disaster declarations stemming from hurricanes have occurred in non-coastal states, such as Arkansas.

Flooding can occur due to multiple factors, including heavy rainfall, snowmelt, storm surges, river overflow, dam failures, and increased urbanization. This study will focus on three main types of flooding: flash floods, river floods, and coastal floods. Flash floods are characterized by rapid increases in water levels that inundate rivers, streams, and roads, typically resulting from intense rainfall or rapid snowmelt. River floods occur when rainfall or snowmelt causes water levels to rise and overflow the banks of the river. Coastal floods typically occur as a result of storm surges during hurricanes. Flash floods can lead to the inundation of ordinarily dry land, exacerbated by both inland and tidal waters, and may also trigger mudslides or the collapse of nearby land (FEMA Title 44 CFR, 2024a).

According to NOAA, a tornado is defined as a "violently rotating column of air that extends from a thunderstorm to the ground" (National Severe Storms Library, 2024). Our study area, known colloquially as the Dixie Valley, is one of the two regions in the United States—alongside the Great Plains—most prone to tornado occurrences (National Geographic Society, 2024). Tornadoes are categorized by intensity on the Enhanced Fujita (EF) Scale, which ranges from EF-0 to EF-5. This scale serves as a tool for evaluating the severity of tornado-related damage where EF-0 indicates minor damage and associated wind speeds of 65–85 miles per hour (approximately 53% frequency) while EF-5 indicates incredible damage and associated wind speeds exceeding 323 miles per hour (approximately 0.05% frequency). While researchers are investigating how changing atmospheric conditions affect the frequency and severity of tornadoes, establishing direct correlations is challenging compared to other extreme weather events. For the purposes of this chapter and our study area, understanding the prevalence of tornadoes and their classifications can aid in preparing healthcare systems and fortifying health facilities alongside essential government and business functions. For example, a comprehensive historical examination of maximum wind speeds can inform the formulation of standards for wind-resistant roofing systems, communications infrastructure, and other structural designs. This understanding, along with the establishment of new building codes and guidelines, has the potential to ensure that health systems remain operational during natural hazard events.

In summary, severe storms—including thunderstorms, tornadoes, and hurricanes—pose significant risks characterized by intense winds, hail, and storm surges. These natural occurrences, along with various causes of flooding including heavy rainfall, snowmelt, and dam failures, greatly impact community infrastructure and health systems (Alderman et al., 2012). Analyzing the classifications and effects of these hazards is central for enhancing preparedness and resilience and ensuring health services remain accessible and available during and after disasters.

Table 5.1 lists the total number of federally declared disasters by incident type within the study area over the last two decades. By examining these statistics, we can identify disaster response and recovery activities and understand the most prevalent types of incidents resulting in natural disasters during the last two decades. Readers should note that flooding is not identified as a primary incident type for Georgia within the examined 20-year period. This suggests that although flooding events have occurred, they were predominantly classified as other primary categories, such as hurricanes or severe storms.

Table 5.1 Leading incident type by state between 2003 and 2023

State	# of events	# of incidents by type			
		Flood	Hurricane	Severe storm	Tornado
Alabama	32	0	10	21	1
Arkansas	30	2	1	23	4
Georgia	18	0	7	10	1
Louisiana	26	6	12	7	1
Mississippi	34	2	8	22	2
Tennessee	34	2	0	29	3

Source: Time of occurrence is based on the incident's begin date. Source: FEMA. OpenFEMA Dataset: Disaster Declarations Summaries - v2 (FEMA, n.d.-d)

Examining the data can help guide resource allocation and prioritization, allowing for targeted support and interventions in areas most frequently affected by specific types of natural disasters. For instance, states along the Gulf Coast (Louisiana, Mississippi, Alabama) experience hurricanes more frequently due to their geographic locations while central states (Tennessee, Arkansas) are more susceptible to severe storms and tornadoes. Specifically, Louisiana is notable for experiencing the most flood incidents, which can affect low-lying regions along the Mississippi River. Severe storms pose a significant threat across most of the states in the study area, particularly in Arkansas and Tennessee, suggesting a regional climatic observance that may require ongoing attention regarding infrastructure and emergency health services. For example, Tennessee recorded a total of 34 incidents with 29 classified as severe storms.

Table 5.2 provides an overview of the total number of counties that received federal disaster declarations across the six-state region from 2003 to 2023 for the four types of incidents: floods, hurricanes, severe storms, and tornadoes. The disaster declarations are based on preliminary damage assessment results that met the criteria for federal assistance. Federal assistance following a disaster can support hazard mitigation strategies during recovery efforts, thereby improving community resilience and reducing risk associated with future events. By examining the data at a geographic scale that is smaller than the state level, we gain valuable insights into the expansive damage resulting from the four incident types across counties in the study area. An evaluation may also examine the recovery measures undertaken to enhance the physical security and resilience of critical facilities. Critical facilities, including public health institutions, are characterized as structures that deliver essential services and functions necessary for ensuring victim survival, sustaining the continuity of public safety operations, and facilitating disaster recovery (FEMA, n.d.-b).

Table 5.2 Top incident types for federal declared major disaster in Alabama, Arkansas, Georgia, Louisiana, Mississippi, and Tennessee 2003–2023

Rank	Incident type	Total # of counties receiving major disaster declarations	Percent
1	Severe storm	1,901	50
2	Hurricane	1,539	41
3	Flood	239	6
4	Tornado	98	3

Note: Percent represents the amount allocated to that category with incident types totaling 100 percent. Time of occurrence is based on the incident begin date. Source: FEMA. OpenFEMA Dataset: Disaster Declarations Summaries - v2 (This product uses the FEMA OpenFEMA API but is not endorsed by FEMA. The Federal Government or FEMA cannot vouch for the data or analyses derived from these data after the data have been retrieved from the Agency's website(s).)

Assessing natural hazards at the county level provides a detailed understanding of their impact on health systems. According to AHRQ's definition from Chapter One, a health system includes at least one hospital and one physician group offering primary and specialty care under unified management (AHRQ, 2023). This integrated care delivery is vital for evaluating health system resilience against natural hazards. To understand community resilience, one should consider the interactions and dependencies among social systems, economic factors, infrastructure, environmental conditions, governance, and housing. This broader view can illuminate the vulnerabilities posed by natural hazards and potential post-disaster scenarios. Furthermore, to facilitate more effective analyses of social-ecological interdependencies and the impacts of adverse weather events, categorizing shocks and stressors is beneficial. Shocks, in this context, are "generally short-duration or acute events that disrupt normal life" while stressors are characterized as "chronic, slow-onset, or long-term conditions that progressively weaken a community, exacerbate the effects of shocks, and hinder overall community functioning and well-being" (FEMA, National Resilience Guidance, 2024b). More specifically, in examining emerging health issues in the context of shocks and stressors, a community can better prepare for and recover from natural disasters.

Consider a flood event where routes (e.g., roads) to a critical health facility are rendered impassable. It is crucial to not only restore the infrastructure to its original condition but also to implement measures to mitigate future flood risks. Specifically, restoration efforts should include effective drainage management or water-channeling strategies to divert water from the road surface. Neglecting such measures can result in recurrent damage and repeated disruptions. Additionally, it is essential to integrate flood risk assessments that utilize historical flooding data alongside predictive models accounting for evolving atmospheric conditions and their impacts on local watersheds and stormwater management systems. Failure to address both existing and potential flood hazards may lead to persistent infrastructure failures and consequent delays in the delivery of critical medical treatments requiring facility-based care. Furthermore, the mitigation of risks to critical infrastructure reduces vulnerabilities and enhances community resilience to environmental shocks. These mitigation measures can prevent infrastructure from damage, thereby sustaining economic activity, ensuring public safety, maintaining critical transportation routes for medical services, and preserving financial resources for community initiatives rather than diverting them to post-disaster recovery effort.

To illustrate, consider the following example. Tattnall County, situated in Georgia's coastal plain approximately 60 miles west of Savannah and 200 miles southeast of Atlanta, presents notable socio-economic and health challenges. The county's per capita income is 43% below the national average, accompanied by elevated prevalence rates of chronic conditions such as chronic obstructive pulmonary disease (COPD), hypertension, and cancer-related mortality (U.S. American Community Survey). Within Tattnall County, the rural communities of Collins and Reidsville are particularly vulnerable to increased flooding risks, resulting from a combination of geological, hydrological, and anthropogenic land development factors that have disrupted natural drainage systems (Anthropocene, 2024). Prolonged exposure to recurrent flooding and its associated stressors can adversely affect physical health, including the proliferation of mold in water-damaged structures. Moreover, the economic and social disruptions from natural disasters contribute to psychological distress, encompassing conditions such as stress, anxiety, and depression. Considering Tattnall County's history of flooding events, planners and healthcare professionals need to integrate these multifaceted vulnerabilities into disaster preparedness and recovery strategies. The implementation of targeted interventions aimed at enhancing stormwater management and reinforcing infrastructure resilience is essential to mitigate environmental shocks and stressors. Furthermore, training first responders and local health practitioners to recognize and address psychological trauma during disaster response and recovery phases is critical. Equipping these professionals with the competencies necessary to identify psychological distress and then deliver timely and appropriate interventions may contribute to improved immediate and long-term health outcomes among affected populations.

Impact on Health Systems

As this chapter was nearing completion, numerous emergency personnel and healthcare professionals were responding to and recovering from system-wide impacts from Hurricanes Helene and Milton. On September 27, 2024, in Tennessee, first responders evacuated 62 patients from Unicoi County Hospital using life rafts and helicopters due to rising floodwater (Feinberg, 2024). In North Carolina, Hurricane Helene caused severe damage to Asheville's water infrastructure, thereby disrupting access to clean drinking water and impairing hospital operations until repairs could be completed (Phillis et al., 2024). Without running water, residents resorted to collecting water from streams for flushing toilets. Some individuals relied on cookstoves and lanterns as alternative sources of light and heat (Kaplan et al., 2024). The storm resulted in extended power outages, disrupted cellular networks and internet services, and obstructed roads and bridges, thereby leaving rural residents without means of communication to stay informed about emergency efforts or to request critical emergency assistance.

When evaluating the impacts of adverse weather events and natural disasters on health systems, it is beneficial to adopt a systems thinking perspective prior to the onset of an emergency event, particularly by analyzing the interconnections among community system components. A valuable framework involves assessing system vulnerability and risk using "community lifelines." The lifelines construct, introduced in 2019 by the U.S. Federal Emergency Management Agency (FEMA), is designed to ensure the continuous operation of essential government and business functions that support public health, safety, and economic security (FEMA, 2025). The framework specifically addresses the following components:

1. **Communications**: Infrastructure for responder communications, public alerts, warnings, and messages, as well as 911 and dispatch services.
2. **Energy:** Power grid and fuel supply necessary to maintain energy services.
3. **Food, Hydration, and Shelter**: Provision of essential resources such as food, hydration, and shelter.
4. **Hazardous Material**: Management of facilities, hazardous materials, pollutants, and contaminants.
5. **Health and Medical**: Medical care, public health services, patient movement, medical supply chains, and fatality management.
6. **Safety and Security**: Law enforcement, fire services, search and rescue, and government services maintaining public safety.
7. **Transportation**: Systems and assets for highways, roadways, mass transit, railways, aviation, and maritime transport.
8. **Water Systems:** Potable water infrastructure and wastewater management.

Source: FEMA, *Community lifelines, 2025.*

An evaluation of Hurricane Helene's impact on Asheville using a community lifelines framework reveals vulnerabilities across multiple systems (e.g., communications, energy, health and medical, transportation, and water systems). Effective planning and preparedness are critical for safeguarding health outcomes and ensuring safety given the interconnectedness of lifeline components and the need for prompt response measures to stabilize health and medical services following a natural disaster. For instance, understanding how individuals with chronic conditions, such as kidney disease, manage their health during such events can substantially inform preparedness strategies, emergency response protocols, and equip patients with critical information necessary for life-preserving interventions. The coordination of lifeline components before, during, and after a disaster is essential for establishing priorities that mitigate risks to human life. Additional recommendations to advance this understanding are discussed in subsequent sections of this chapter.

Emerging Health Considerations

While community planning and preparedness are essential for enhancing resilience, the question remains regarding the capacity of these systems to adapt to changing atmospheric conditions and the effects of adverse weather events. Given the compounded impacts of adverse weather events, disasters, and the public health sector's dependence on complex systems, we must adopt a novel approach to public health outcomes. This includes examining individual and community exposure to shocks and stressors, as well as incorporating community resilience within the lifelines construct. Additionally, we need to address the decision-making and policy challenges related to improving the adaptive capacity of socio-ecological systems (Vallury et al., 2022).[1] The primary question being, can these systems adapt to compound disasters and adverse weather events? Consider the following emerging health issues:

1. **Cardiovascular Health:** Research indicates a significant increase in cardiovascular hospitalizations associated with disaster events. For instance, a study of post-Katrina cardiovascular disease (CVD) hospitalization rates in Louisiana found an immediate rise in CVD hospitalizations following Hurricane Katrina's landfall, a trend that continued for a month (Becquart et al., 2019).
2. **Drinking Water and Sanitation:** Flooding poses serious risks to public health by contaminating drinking water facilities and creating stagnant water sources that can serve as breeding grounds for mosquitoes. Additionally, the presence of floodwater introduces chemical hazards and increases the risk of injuries. Similar to other natural disasters, flooding disrupts health systems, facilities, and services, thereby limiting community access to healthcare and safe drinking water. Moreover, flooding damages essential infrastructure, including water supply and transportation networks, which can delay emergency response efforts and compromise sanitation practices for healthcare facilities dependent on reliable power and water sources.
3. **Food Insecurity/Healthy Food:** Access to healthy food and clean water becomes scarce, forcing communities to rely on preserved foods that do not require refrigeration. Many of these preserved foods are high in sodium and sugar, exacerbating pre-existing conditions such as cardiovascular disease and diabetes.
4. **Healthcare Services:** Regular care facilities may be closed or destroyed during major disasters, disrupting health care services. Transportation routes and methods may be impacted. This issue is particularly concerning in rural communities that already are challenged by limited health offerings.
5. **Health Equipment and Supplies:** Looking beyond the community to the broader supply chain, health systems should ensure multiple supplier agreements are in place to support manufacturing disruption due to extreme weather events. In the aftermath of Hurricane Helene, hospitals across the country experienced shortages of IV fluid bags after Baxter International, a key supplier, was forced to close one of its IV manufacturing plants due to the hurricane's devastating effects (HHS, 2024).

---

[1] Based on my experiences and readings, I will put forth a general definition of adaptive capacity as "the ability to enhance resilience in response to change."

6. **Social Connections:** Disasters have the potential to separate families and cause networks to be disrupted due to casualties, relocation, or destruction of facilities. Moreover, stress induced by disasters can influence relationships, potentially affecting both emotional support and resource-based assistance and may lead to social isolation. Social isolation can have detrimental health effects, including an increased risk of depression, heart disease, a weakened immune system, among other negative outcomes.
7. **Stress Impacts:** The cascading impacts of adverse weather events are linked to an increased demand for mental health services. A recent study on the compounded effects of hurricanes revealed that "repeated, direct, indirect, and media exposures" correlate with increased psychological stress and post-traumatic stress disorder or PTSD (Garfin et al., 2022). Notably, several states within the study area rank among the highest for prevalence of mental health conditions (CDC, 2024). Additionally, substantial evidence connects prolonged stress to various health issues, including cardiovascular health issues and obesity.
8. **Telemedicine:** Expanding on the rising need to support emerging health issues in rural areas following single or multiple disaster events is the utilization of telemedicine. For instance, rural Tennesseans often encounter barriers such as transportation, broadband, cell service availability, and platform knowledge when accessing provider-based or in-home telemedicine services (Tennessee Rural Health Taskforce, 2023). In designing this access, a more adaptable approach would involve ensuring that telemedicine infrastructure is capable of withstanding severe wind intensity and providing redundancy in internet delivery to prevent service interruptions.

These examples highlight some of the public health challenges resulting from adverse weather events and natural disasters. A thorough understanding of these issues can enhance preparedness initiatives and result in improved disaster recovery outcomes, particularly when determining how to prioritize limited health and medical resources. Considering emerging health issues from individual, organizational, and community perspectives is useful when outlining steps for the adoption of new policies and behaviors needed to address challenges. While various models delineate steps for successful change and adoption of new policies and behaviors, nearly all begin with awareness.

Flood Awareness and Stressors

Hurricanes and flooding events are relatively common in the study area, often necessitating evacuations. Residents in the evacuation zones typically seek shelter with friends or family, relocate to designated facilities, or choose to stay in their homes. The duration of displacement can vary widely, lasting from a few days to several months. This sudden increase in temporary population may strain healthcare systems, particularly in rural areas where resources are already limited (see Chapter

Two). Understanding these dynamics can reveal the cascading impacts of adverse weather events that extend beyond jurisdictional boundaries. The displacement of local populations underscores emerging public health challenges, notably regarding insurance coverage, access to pharmaceuticals, state aid, and adequate support for healthcare facilities and staffing.

For example, by considering the specific needs of older adults, negative health impacts may be reduced when providing temporary housing following a disaster. Research suggests that older residents who remain in their homes within their community often experience improved mental health outcomes and decreased feelings of loneliness due to neighborhood features such as accessibility and social cohesion (Van Dijk et al., 2023). Temporary housing can potentially disrupt this by moving individuals from independent living to alternative accommodations. Addressing the needs of older adults during recovery and considering factors, such as access to essential facilities and the level of social connectedness when evaluating temporary housing options, can help lessen the emotional distress for those who are temporarily displaced and relocated.

It is essential to consider persistent stressors associated with residing in hazard-prone areas, particularly those related to housing financial concerns, including insurance costs, availability, and coverage. Post-disaster flood studies frequently inform updated flood modeling, which can influence actuarial risk ratings and subsequently result in modifications to property risk assessments and insurance premiums. An increase in premiums will disproportionately impact older adults on fixed incomes, who may have difficulty maintaining insurance affordability. This challenge is further compounded when the ability to sell property is constrained as prospective buyers may be deterred by elevated insurance costs alongside existing flood risks. Consequently, affected individuals may encounter obstacles in realizing home equity, a key component of their financial security. The psychological and economic strains experienced by residents during natural disasters can be exacerbated when available insurance coverage, despite being affordable, does not adequately compensate for replacement costs. Given escalating disaster-related expenses and insurance premiums, alongside the pervasive fear of permanent displacement in communities experiencing recurrent and compounded hazards, public health professionals need to recognize these factors as chronic stressors. Effective intervention requires the integration of targeted counseling and resource provision within patient care services to enhance mental health outcomes and overall well-being among affected populations.

Meeting the health needs of residents contributes to stability and security by enabling a quicker return to routine activities and the restoration of community services, thereby improving the community's adaptive capacity. Additionally, by considering stressors in relation to health system impacts and natural hazard risks, public health professionals can create a valuable framework for identifying emerging health issues related to adverse weather scenarios. To understand community resilience, healthcare practitioners and policymakers should consider the interactions and dependencies among social systems, economic factors, infrastructure, environmental conditions, governance, and housing (FEMA, Aug. 2024). This broader view can illuminate the vulnerabilities posed by natural hazards and inform planning around hypothetical "what if" scenarios.

Community-based Resilience

The previous chapters highlighted socio-economic challenges in rural areas, such as outdated infrastructure, limited internet access, a shortage of healthcare professionals, extreme poverty, and restricted educational opportunities. Specific health issues prevalent in the study area, including cardiovascular disease, diabetes, and obesity, were also detailed. By considering these stressors in the context of health system impacts and known natural hazard risks, we can create a valuable framework for identifying emerging health issues associated with adverse weather scenarios. Recognizing that economically disadvantaged populations are disproportionately impacted by disasters highlights the need to expand on the system-based approach and the utilization of lifelines with a cultural dimension. This perspective emphasizes not just health outcomes but also the resilience of the entire community. The National Institute of Standards and Technology (NIST) defines community resilience as the ability to prepare for anticipated hazards, adapt to changing conditions, and swiftly recover from disruptions (NIST, 2024). For rural health practitioners and public service professionals, understanding community resilience can assist with better planning and response so health systems can withstand and recover more quickly from natural disasters. Additionally, understanding community resilience assessments can help identify vulnerable populations while also enhancing social cohesion and support, which are vital for mental and physical health during disaster events (SAMHSA, 2017).

One essential resource that supports this assessment is the CDC/ATSDR Social Vulnerability Index (SVI). It utilizes 16 census variables from the five-year American Community Survey (ACS). These variables are categorized into four major themes of social vulnerability, resulting in a composite social vulnerability score. The following Table 5.3 summarizes these four themes in the first row with the corresponding 16 variables in the second row (see https://www.atsdr.cdc.gov/place-health/php/svi/index.html). Another resource to consider is the National Risk Index (NRI) dataset, which categorizes 18 different hazard types. It can be utilized to enhance public awareness about the specific risks present in a community (refer to Table 5.4 and see https://hazards.fema.gov/nri/community-resilience).

Table 5.3 Social vulnerability index (CDC/ATSDR SVI) four themes and related variables used to determine SVI

Socio-economic status	Household characteristics	Racial and ethnic minority status	Housing type and transportation
Below 150% poverty Unemployed Housing cost burden No high school diploma No health insurance	Aged 65 and older Aged 17 and younger Civilian with a disability Single-parent households English language proficiency	Hispanic or Latino (of any race) Black and African American, Not Hispanic or Latino American Indian and Alaska Native, not Hispanic or Latino Asian, not Hispanic or Latino Native Hawaiian and Other Pacific Islander, Not Hispanic or Latino Two or more races, not Hispanic or Latino Other races, not Hispanic or Latino	Multi-unit structures Mobile homes Crowding No vehicle Group quarters

Table 5.4 National risk index: 18 hazard types

Avalanche	Heat wave	Strong wind
Coastal flooding	Hurricane	Tornado
Cold wave	Ice storm	Tsunami
Drought	Landslide	Volcanic activity
Earthquake	Lightening	Wildfire
Hail	Riverine flooding	Winter weather

Source: FEMA National Risk Index Technical Documentation (https://www.fema.gov/sites/default/files/documents/fema_national-risk-index_technical-documentation.pdf) (FEMA, 2023).

Application of the Study Area

Rural counties frequently lack the infrastructure and equipment necessary to provide dialysis services, often possessing only a single dialysis facility, if any (Crouch et al., 2024). As highlighted in the previous chapter, the widespread occurrence of diabetes in the study area is concerning, particularly since individuals with diabetes may necessitate dialysis. Diabetes (e.g., diabetic kidney disease) is a leading cause of kidney failure (de Boer et al., 2022). In 2022, states within the study area recorded some of the highest rates of kidney disease mortality in the United States with age-adjusted death rates ranging from 17.5 to nearly 22.6 per 100,000 total population (CDC, 2022).

While some patients are able to receive home dialysis treatments, such as home hemodialysis or peritoneal dialysis, these methods depend on a stable supply of electricity and necessary supplies. Disasters can disrupt supply chain production and delivery, as well as cause power outages. Access to kidney dialysis can become critically limited during emergencies or natural disasters due to extended power failures and road closures. This section will demonstrate the application of the tools and methodologies presented in this chapter by examining the specific needs of individuals with kidney disease and the challenges associated with interruptions to dialysis treatment resulting from adverse weather events.

Kidney Disease and Dialysis in Rural Communities in the Six-State Study Area

The National Institute of Diabetes and Digestive and Kidney Diseases reports that over 800,000 individuals in the United States are affected by end-stage renal disease (ESRD) with 68% reliant on dialysis. ESRD is more common among individuals aged 65 and older. According to the American Kidney Fund (n.d.), Black Americans comprise approximately one-third of the population diagnosed with kidney failure in the United States. Chronic kidney disease (CKD), which frequently progresses to ESRD, demonstrates the highest prevalence among individuals aged 65 and older with a reported rate of 34%. Furthermore, by the conclusion of 2021, 14.2% of patients undergoing dialysis received treatment in a home setting, reflecting a 45% increase in-home dialysis utilization since 2011 (NIDDKH, 2024).

To understand the dialysis requirements associated with CKD and ESRD within a specific geographical context, a range of risk factors must be considered. These include family history, advanced age, comorbid conditions such as diabetes, hypertension, and cardiovascular disease, as well as ethnicity and lifestyle factors. Epidemiological evidence consistently demonstrates a higher prevalence of CKD among older adults. For the purpose of this analysis, the focus will be on Natchitoches, Louisiana, a city located in the northwest region of the state. Established in 1714, Natchitoches is recognized as the oldest permanent settlement within the Louisiana Purchase territory (World Atlas, n.d.).

To enhance our understanding of this area, we can collect additional information regarding previously identified risk factors of CKD. As of 2023, Natchitoches has an estimated population of 17,600. The demographic composition consists of approximately 56% identifying as Black or African American and 35% as White. The median age in Natchitoches is 26.7 years with 17% of the population aged 60 years and older. The city struggles economically with a poverty rate of 34.8%, which is nearly double the state rate of 18.9%. The median household income is reported at $38,731 while the average household income is reported at $60,359 (U.S. Census Bureau, n.d.). Health surveillance data from the Louisiana Department of Health indicate a diabetes prevalence of 10.4% among residents of Natchitoches Parish for 2021 (Louisiana Dept. of Health, n.d-a). Meanwhile, recent estimates from the Behavioral Risk Factor Surveillance System (BFRSS) show an increase with a prevalence of 15% for 2025 (UW PHI, 2025). Rural areas such as Natchitoches face challenges including limited access to healthcare facilities and resources, compounded by escalating substance abuse. State-level epidemiological data reveal a considerable increase in opioid-related morbidity and mortality, predominantly attributed to fentanyl. In 2023, Louisiana reported 2,456 drug-related fatalities with a significant proportion, approximately 1,130 cases, associated with opioid consumption. Additionally, there were 13,937 emergency department visits attributed to drug poisoning during the same period (Louisiana Dept. of Health, n.d-b). These factors collectively emphasize the complex interplay of socio-economic and health determinants impacting the region.

By examining the Social Vulnerability Index (SVI) alongside statewide socio-economic and minority status indicators, health professionals can assess the susceptibility of various social groups to natural hazard events (see Table 5.3). For example, certain census tracts within Natchitoches Parish are classified within the "highest top 4th quantile" of overall social vulnerability, as depicted in Fig. 5.1. Social vulnerability is defined as a community's capacity to prepare for and respond to the stresses associated with hazardous incidents (CDC/ATSDR, SVI, 2022). It can also be considered as "the susceptibility of social groups to the adverse effects of natural hazards, including disproportionate death, injury, loss, or disruption of livelihood" (FEMA, 2023).

Natchitoches Parish has a "relatively low" **overall risk index**. However, the **social vulnerability** is assessed as "very high" and **community resilience** as "relatively low." This suggests that certain social groups within the Parish exhibit a heightened susceptibility to the adverse effects of natural hazards relative to national averages (FEMA, n.d.-f and see https://www.fema.gov/emergency-managers/practitioners/resilience-analysis-and-planning-tool). For instance, some communities in the Parish may experience comparatively limited capacity to prepare for natural hazards, adapt to changing conditions, and recover from disruptions. This emphasizes the importance of considering social vulnerabilities when planning actions and activities to bolster public health preparedness. Readers should note that while this tool provides useful information to inform community planning approaches, a comprehensive understanding of natural hazard risk can be obtained by reviewing the communities hazard mitigation plan. Integrating this knowledge will facilitate targeted strategies aimed at enhancing the stability of critical lifelines, strengthening community resilience, and ultimately improving public health outcomes during and after adverse weather events.

Healthcare professionals can assess the risk of CKD within the context of natural disasters by examining the dependence of local populations on kidney dialysis, anticipating potential hazard scenarios impacting the community, and proactively developing and implementing response protocols. Obtaining a comprehensive understanding of natural hazard risks within a community necessitates interdisciplinary collaboration among healthcare providers, emergency management personnel, and public planning authorities. For example, the 2023 Natchitoches Multi-Jurisdictional Hazard Mitigation Plan categorizes adverse weather events

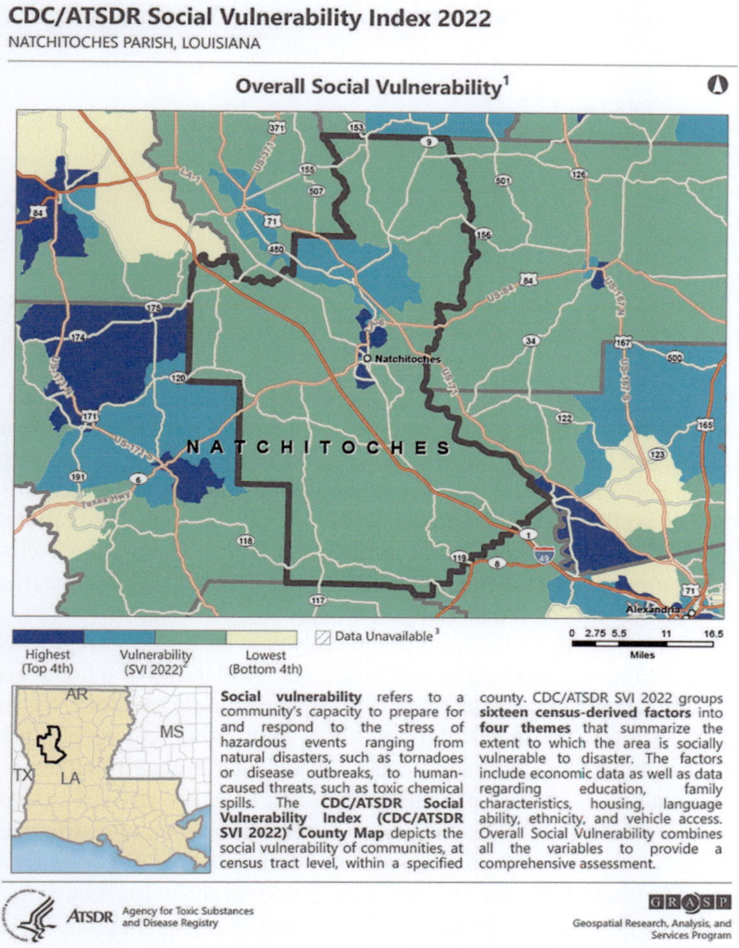

Fig. 5.1 SVI for Natchitoches Parish, LA. Overall score of 0.78 (i.e., high vulnerability). (Source: CDC SVI tool)

as high-risk hazards when the overall risk score is 2.5 or greater on a scale from 1 to 4 (Table 5.5). This risk factor was determined by applying a weighted evaluation system, wherein values for probability, impact, spatial extent, warning time, and duration were multiplied by corresponding weighting factors and subsequently aggregated to yield a composite risk score. See equation:

Risk Factor = (Probability × 0.25) + (Impact × 0.25) + (Spatial Extent × 0.20) (Warning Time × 0.15) + (Duration × 0.15)

Table 5.5 Risk factors for different hazards

Hazard	Overall risk
Drought	2.8
Flooding	3.4
Thunderstorms	2.2
Thunderstorms—winds	2.7
Tornadoes	2.9
Tropical cyclones	3.3
Wildfires	3.2

Adapted from Table 2.9-Risk Assessment for Natchitoches Parish Natchitoches Parish, Natchitoches Parish Hazard Mitigation Plan (SDMI, 2023)

Analyzing the hazards listed in the table and considering how they impact patient needs, such as dialysis dependency, can provide a critical perspective for communities planning and preparing for emergencies. Individuals with a comprehensive understanding of risk factors associated with both existing and emerging health conditions, as well as specific patient needs, are encouraged to collaborate with representatives across various lifeline components. This collaboration is crucial for enhancing community resilience and addressing public health concerns when adverse weather events disrupt essential services.

For instance, patients utilizing in-home dialysis who reside in regions with an elevated risk of natural disasters should ensure the availability of a home generator and maintain an adequate reserve of intravenous fluids and dialysis supplies. Additionally, patients must be informed about appropriate emergency protocols, including the preparation of healthy meals during such events. Healthcare professionals serving kidney disease patients in high-risk areas are advised to implement a "3 Day Emergency Diet." This dietary protocol is structured to restrict the intake of potassium, protein, and fluids in instances where dialysis sessions are limited (Center for Medicare and Medicaid Services, 2015). In scenarios requiring evacuation or temporary shelter, public health officials should engage in proactive collaboration with local emergency management services to ensure that temporary housing and shelter arrangements are situated close to alternative dialysis centers. Concurrently, systematic tracking of the number and geographic distribution of affected patients should be undertaken with strict adherence to patient confidentiality protocols. The establishment of data-sharing agreements prior to disaster events is also recommended to facilitate the timely exchange of vital public health information. Additionally, registration with utility companies may facilitate the priority restoration of essential services, such as electricity and water, following a disaster event.

The challenges faced by rural counties in providing adequate dialysis services demonstrate the need for comprehensive disaster preparedness and management strategies. Ensuring reliable access to electrical power and necessary supplies for home-based dialysis, alongside planning for emergency situations, is essential for safeguarding the health and well-being of patients with kidney disease. Healthcare providers and public health authorities are encouraged to collaborate proactively to mitigate the impact of disasters on dialysis services, including implementing necessary infrastructure modifications and promoting awareness among patients and their caregivers. By prioritizing these initiatives, it is possible to strengthen health system resilience and improve patient outcomes in the aftermath of natural disasters and other emergency events.

Before concluding, I offer an additional valuable resource for evaluating community resilience: the Resilience Analysis Planning Tool (RAPT). This comprehensive geographic information system (GIS) integrates multiple data layers related to infrastructure, hazards, and community demographics. More information is available on FEMA's website with supporting documents (visit https://www.fema.gov/emergency-managers/practitioners/resilience-analysis-and-planning-tool or search for FEMA RAPT). For healthcare professionals, RAPT is particularly beneficial as it offers essential health infrastructure data—such as locations of assisted living facilities, dialysis centers, hospitals, and urgent care centers—allowing for a clear visualization of community resources. Users can explore the data at the county, census tract, or municipal level, and they also have the option to integrate their own data. This user-friendly resource is free to access and is worth exploring.

Conclusion

The increasing frequency and severity of weather-related disasters across Alabama, Arkansas, Georgia, Louisiana, Mississippi, and Tennessee highlight the urgent need for strong community resilience and well-prepared health systems. While forecasting models for hurricanes, severe storms, and tornadoes have improved, a degree of uncertainty persists. This inherent unpredictability necessitates the development and implementation of comprehensive disaster preparedness and response strategies to reduce health risks associated with floods, hurricanes, and other natural hazards.

Expanding community health frameworks to integrate preparedness, response, and recovery is critical to protecting public health and enhancing overall well-being. The compounded impacts posed by severe storms, hurricanes, floods, and tornadoes necessitate a system-thinking approach that integrates social, economic, and environmental factors to bolster adaptive capacity and resilience. In the context of rural communities, healthcare professionals need to focus on several key strategies to enhance the delivery of essential health services. These strategies include the promotion of preventive health measures, the enhancement of telemedicine infrastructure, the assurance of reliable access to essential medical supplies, and the cultivation of collaborative partnerships among healthcare providers and emergency management services. By implementing these measures, rural residents can better access health services, thereby improving health outcomes and quality of life within these communities. Through effective management and resilient emergency response systems, we can better protect vulnerable populations and reduce health disparities during natural disasters.

References

Agency for Healthcare Research and Quality, AHRQ. (2023). *Defining health systems.* AHRQ. https://www.ahrq.gov/chsp/defining-health-systems/index.html

Alderman, K., Turner, L. R., & Tong, S. (2012). Floods and human health: a systematic review. *Environment International, 47*, 37–47.

American Kidney Fund. (n.d.). *Understanding kidney disease risks: Race and ethnicity.* American Kidney Fund. https://www.kidneyfund.org/all-about-kidneys/risk-factors/understanding-kidney-disease-risks-race-and-ethnicity.

Anthropocene Alliance. (2024). *Reidsville/Collins restorative landscape plan: Existing conditions report.* https://anthropocenealliance.org/wp-content/uploads/2024/01/Reidsville_Collins-Restorative-Landscape-Plan-Ex-Conditions-Report.pdf

Becquart, N. A., Naumova, E. N., Singh, G., & Chui, K. K. (2019). Cardiovascular disease hospitalizations in Louisiana parishes' elderly before, during and after hurricane Katrina. *International Journal of Environmental Research and Public Health, 16*(1), 74.

Centers for Disease Control and Prevention (CDC). (2022). *Stats of the States—Kidney disease mortality.* https://www.cdc.gov/nchs/pressroom/sosmap/kidney_disease_mortality/kidney_disease.htm

Centers for Disease Control and Prevention (CDC). (2024). *Preparing for the regional health impacts of climate change in the United States: A summary of health effects, resources, and adaptation examples from health departments funded by CDC's Climate and Health Program.* https://www.cdc.gov/climate-health/php/climate_ready/index.html

Centers for Disease Control and Prevention (CDC) & Agency for Toxic Substances and Disease Registry. (2022). *Social Vulnerability Index (SVI).* https://www.atsdr.cdc.gov/placeandhealth/svi/index.html

Centers for Medicare & Medicaid Services. (2015). *Preparing for emergencies: A guide for people on dialysis.*

Crouch, E., Yell, N., Herbert, L., Browne, T., & Hung, P. (2024). Availability and quality of dialysis care in rural versus urban US counties. *American Journal of Nephrology, 55*(3), 361–368.

de Boer, I. H., Khunti, K., Sadusky, T., Tuttle, K. R., Neumiller, J. J., Rhee, C. M., & Bakris, G. (2022). Diabetes management in chronic kidney disease: a consensus report by the American Diabetes Association (ADA) and Kidney Disease: Improving Global Outcomes (KDIGO). *Diabetes Care, 45*(12), 3075–3090.

Federal Emergency Management Agency (FEMA). (2023). *National risk index technical documentation.* Federal Emergency Management Agency (FEMA). https://www.fema.gov/sites/default/files/documents/fema_national-risk-index_technical-documentation.pdf

Federal Emergency Management Agency (FEMA). (2024a). *Code of Federal Regulations (CFR) Title 44: Emergency management and assistance.* https://www.ecfr.gov/current/title-44

Federal Emergency Management Agency (FEMA). (2024b). *National resilience guidance: A collaborative approach to building resilience.* Federal Emergency Management Agency (FEMA). https://www.fema.gov/emergency-managers/national-preparedness/plan/resilience-guidance

Federal Emergency Management Agency (FEMA). (2025). *Community lifelines.* FEMA. Retrieved April 13, 2025, from https://www.fema.gov/emergency-managers/practitioners/lifelines

Federal Emergency Management Agency (FEMA). (n.d.-a). *How a disaster gets declared.* U.S.. https://www.fema.gov/disaster/how-declared

Federal Emergency Management Agency (FEMA). (n.d.-b). *IS-0815: The A-B-Cs of temporary emergency Power—Lesson 1: Course introduction and critical facility identification.* Federal Emergency Management Agency (FEMA). Retrieved April 11, 2025, from https://emilms.fema.gov/IS0815/curriculum/1.html

Federal Emergency Management Agency (FEMA). (n.d.-d). *OpenFEMA Dataset: Disaster Declarations Summaries-v2.* Federal Emergency Management Agency (FEMA). https://www.fema.gov/api/open/v2/DisasterDeclarationsSummaries

Federal Emergency Management Agency (FEMA). (n.d.-f). National risk index, risk comparison report. Retrieved and produced on April 20, 2025, from https://hazards.fema.gov/nri/

Feinberg, A. (2024, September 30). "We were on an Island": First responders recount Unicoi County Hospital rescue mission. *Knox News, The Knoxville News-Sentinel*. https://www.knoxnews.com/story/news/local/2024/09/28/tennessee-unicoi-county-hospital-rooftop-rescue-floods-first-responders/75427020007/

Garfin, D. R., Thompson, R. R., Holman, E. A., Wong-Parodi, G., & Silver, R. C. (2022). Association between repeated exposure to hurricanes and mental health in a representative sample of Florida residents. *JAMA Network Open, 5*(6), e2217251–e2217251.

Kaplan, S., Dennis, B., & Dance, S. (2024, October 9). *How Asheville residents survive without running water, weeks after Helene.*

Lee, H., Calvin, K., Dasgupta, D., Krinner, G., Mukherji, A., Thorne, P., & Zommers, Z. (2023). *Climate change 2023: synthesis report. Contribution of working groups I, II and III to the sixth assessment report of the intergovernmental panel on climate change*. The Australian National University.

Louisiana Department of Health. (n.d.-a). *Health data explorer*. Louisiana Department of Health. https://healthdata.ldh.la.gov/

Louisiana Department of Health. (n.d.-b). *Louisiana opioid data and surveillance system*. Louisiana Department of Health. https://lodss.ldh.la.gov/

National Geographic Society. (2024). Tornadoes. *National Geographic*. https://education.nationalgeographic.org/resource/tornadoes-and-enhanced-fujita-scale/

National Institute of Diabetes and Digestive Kidney Diseases (NIDDKH). (2024). *Kidney disease statistics for the United States*. National Institutes of Health. https://www.niddk.nih.gov/health-information/health-statistics/kidney-disease

National Institute of Standards and Technology (NIST). (2024). *Community resilience*. U.S. Department of Commerce. https://www.nist.gov/community-resilience

National Oceanic and Atmospheric Administration (NOAA). (2024). *What is a hurricane?*. https://oceanservice.noaa.gov/facts/hurricane.html

National Oceanic and Atmospheric Administration (NOAA). (n.d.). *What are severe storms?*. http://www.weather.gov/

National Severe Storms Laboratory. (2024). *Weather briefly: Tornadoes*. NOAA. https://www.nssl.noaa.gov/education/svrwx101/tornadoes/

Phillis, M., Amy, J., & Peterson, B. (2024, October 4). *A week after Helene cut water supplies, thousands in North Carolina struggle to find enough*. PBS NewsHour https://www.pbs.org/newshour/nation/a-week-after-helene-cut-water-supplies-thousands-in-north-carolina-struggle-to-find-enough

Stephenson Disaster Management Institute (SDMI). (2023). *Natchitoches Parish Multi-Jurisdictional Hazard Mitigation Plan Update*. Louisiana State University. https://hmplans.sdmi.lsu.edu/Home/Parish/natchitoches

Substance Abuse and Mental Health Services Administration (SAMHSA). (2017). *Disaster technical assistance center supplemental research bulletin: Greater Impact: How disasters affect people of low socioeconomic status*. Substance Abuse and Mental Health Services Administration (SAMHSA). https://www.samhsa.gov/sites/default/files/dtac/srb-low-ses_2.pdf

Tennessee Rural Health Care Task Force. (2023). *Improving rural health care across Tennessee*. Tennessee Rural Health Care Task Force. https://www.tn.gov/content/dam/tn/health/program-areas/rural-health/Final-TN-Rural-Health-Care-Task-Force-Report-6-27-23.pdf

U.S. Census Bureau. (n.d.). *Census Reporter for Natchitoches, LA*. U.S. Census Bureau. https://censusreporter.org/profiles/16000US2253545-natchitoches-la/

U.S. Department of Health and Human Services (HHS). (2024). *Fact Sheet: HHS continues taking action to increase access and supply of IV fluids following Hurricane Helene*. U.S. Department of Health and Human Services. https://www.hhs.gov/about/news/2024/10/18/fact-sheet-hhs-continues-action-increase-access-supply-iv-fluids-hurricane-helene.html

University of Wisconsin Population Health Institute (UW PHI). (2025). *Diabetes prevalence*. County Health Rankings & Roadmaps. https://www.countyhealthrankings.org/health-data/population-health-and-well-being/quality-of-life/physical-health/diabetes-prevalence

Vallury, S., Smith, A. P., Chaffin, B. C., Nesbitt, H. K., Lohani, S., Gulab, S., & Allen, C. R. (2022). Adaptive capacity beyond the household: a systematic review of empirical social-ecological research. *Environmental Research Letters, 17*(6), 063001.

Van Dijk, H. M., Cramm, J. M., & Nieboer, A. P. (2023). Neighbourhood qualities are related to better mental health, quality of life, and reduced loneliness among older adults. *The Gerontologist, 65*, gnaf095. https://doi.org/10.1093/geront/gnaf095

World Atlas. (n.d.). *The oldest settlement of the Louisiana Purchase*. World Atlas. https://www.worldatlas.com/destination/the-oldest-settlement-of-the-louisiana-purchase.html

Chapter 6
Conclusion

Abstract This concluding chapter examines the effects of rising temperatures on rural communities with primary emphasis on health implications. Additionally, it addresses the broader effects on agriculture, economic stability, community dynamics, and overall quality of life. An illustrative example is found in Northeast Arkansas, a predominantly rural region that has experienced a temperature increase ranging from 0.7 to 1.2 degrees Fahrenheit over the past century. This example provides a contextual framework for understanding these issues. A supplementary analysis of heat stress in Southeast Louisiana, including Parishes near Baton Rouge and New Orleans, is included to illustrate regional variations in temperature impacts. Demographic factors, including the aging rural population and patterns of rural out-migration, are also examined in relation to population health policies. The chapter concludes by providing a collection of additional resources intended to support initiatives focused on improving health outcomes within rural populations.

Keywords Heat · Rural communities · Rural health · Rural population

Overview

An interdisciplinary approach is essential for understanding both current and emerging rural health issues in the Southeast United States. Such a framework facilitates a meaningful exploration of historical and socio-cultural factors that shape the health landscape in this region. For example, my analysis of the westward expansion of early settlers highlights the growth of towns and the subsequent relocation of Indigenous populations from their ancestral territories—these lands were not merely geographical locations but integral parts of Indigenous cultural, religious, and social identities. The establishment of land reservations granted Indigenous communities a degree of self-governance and autonomy. However, it simultaneously resulted in social isolation. As discussed in Chap. 2, this historical context combined with an understanding of early healthcare practices, rural health

A. Sharma, *Current and Emerging Rural Health Issues in the Southeast United States*, Global Perspectives on Health Geography,
https://doi.org/10.1007/978-3-032-07543-7_6

disparities, and the evolution of healthcare delivery systems provides context amidst a transformative period in American history. In the subsequent chapter, I expanded upon foundational concepts by examining social and racial considerations influencing health outcomes among marginalized populations, including Indigenous groups, African Americans, women, and economically disadvantaged White males. For example, the women's suffrage movement emphasized the importance of policies that addressed maternal and infant health, improved nutrition, and ensured safe working conditions during a period of rapid societal change. Effectively addressing the prevalent health issues in the Southeast, such as coronary heart disease, diabetes, and obesity, requires a nuanced understanding of these historical issues and social determinants of health.

Adverse weather events complicate the healthcare landscape in this area, exacerbating existing health issues and introducing new challenges to healthcare delivery. This region is particularly vulnerable due to its unique geographic and climatic characteristics which encompass changing precipitation patterns, increasing temperatures, and severe weather events. Such adverse weather changes have implications for public health, particularly among communities with limited resources. This situation is compounded by the increasing prevalence of chronic diseases, such as coronary heart disease and diabetes (Vaidyanathan, 2024). The region's agricultural dependence adds another layer of complexity to the healthcare landscape (see Chap. 1). Variability in weather patterns and the occurrence of adverse events can disrupt food production, leading to food insecurity, which is a critical social determinant of health (Ingram et al., 2013). Consequently, residents may be compelled to rely on processed, high-calorie foods that are more readily available but are low in nutrition. In areas where healthy food options are limited, residents may develop poor dietary habits. This transition can increase the risk of developing chronic diseases. This could strain already challenged healthcare systems. While the previous chapter examined several important health challenges related to adverse weather and natural disasters, this discussion will explore an additional related factor before concluding.

Emerging Health Issues: Adverse Weather Events

Heat

We should recognize that weather patterns are shaped by a myriad of factors, one of which is a steady rise in global temperatures. This increase in global temperatures, while a significant concern, is also influenced by an array of elements including human activities, natural phenomena, and even geological processes (Masson-Delmotte et al., 2019). Although this book does not delve into the intricate causal mechanisms behind these changes, readers should note that the Southeast United States has been experiencing—and will continue to experience—weather patterns that feature prolonged summer seasons marked by elevated temperatures and spring seasons characterized by increased precipitation.

To illustrate this trend, Fig. 6.1 highlights the rate of temperature change in the study area from 1901 to 2021. Created by the Office of Air and Radiation within the Environmental Protection Agency (EPA), this map is based on a comprehensive analysis of various atmospheric indicators. Before we explore the implications of this map further, we should clarify two key terms.

Weather describes the atmospheric conditions at a particular place and time. Most weather phenomena that impact human activities, agriculture, and ecosystems occur within the lower layers of the atmosphere. This makes weather a fundamental aspect of our daily lives with changes possible in a matter of minutes or hours.

In contrast, climate refers to the long-term patterns and averages of meteorological conditions observed within a specific geographical region. While weather is often inconsistent and can vary drastically over short periods, climate evolves slowly, typically taking decades to centuries to manifest noticeable changes (U.S. EPA, 2024).

Understanding the distinction between these two concepts is crucial for grasping the wider environmental changes impacting both the Southeast United States and the global ecosystem.

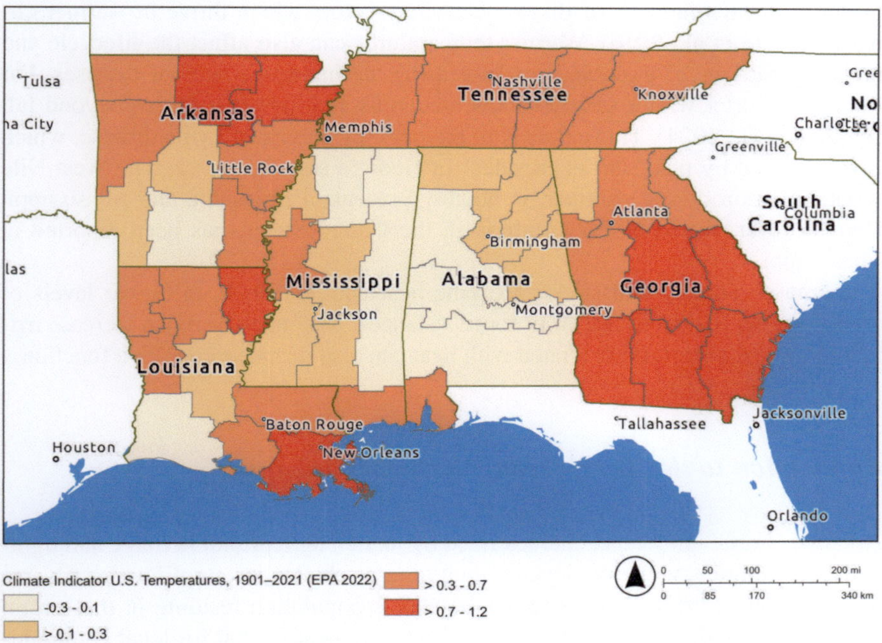

Fig. 6.1 Temperature changes from 1901 to 2021 (°F). (Source: EPA Office of Air & Radiation, Office of Atmospheric Programs, Office of Mission Support, ESRI, TomTom, Garmin, FAO, NOAA, USGS, EPA, USFWS (A. Sharma))

This choropleth map vividly illustrates the geographical areas where temperatures, recorded in Fahrenheit, have risen over the past century. Among the regions highlighted, particular attention is drawn to the northeast section of Arkansas, the far southeastern tip of Louisiana, and the majority of central and southern Georgia. The long-term temperature trend in these areas indicates an increase ranging from 0.7 °F to 1.2 °F. The observed numerical change, while seemingly marginal, reflects an approximate rise of 3 °F in the frequency of extreme hot days. This increase translates to nearly two additional months each year during which temperatures exceed 95 °F in the study area. This level of warming poses a range of potential health-related issues.

Heat exhaustion and heat stroke: Prolonged exposure to elevated temperatures can lead to heat-related illnesses (Heidari et al., 2020). Heat exhaustion, characterized by heavy sweating, weakness, and fainting, can develop into heat stroke. Heat stroke is a critical medical condition where the body's thermoregulatory mechanism no longer functions effectively.

Increased cardiovascular issues: Higher temperatures can lead to cardiovascular straining because the body exerts more energy and effort to cool itself, particularly in cases with obesity. This can elevate heart rates and blood pressure, potentially leading to heart attacks or other cardiovascular events (Kuehn, 2023).

Extended habitats for vectors with greater transmission: Higher temperatures can expand the environments of disease-carrying vectors which thrive in warmer climates (Beard et al., 2016). Warmer temperatures can also affect the lifecycle and breeding patterns of these vectors, increasing the incidence rates of diseases. For example, shorter winter seasons allow arachnids and insects to thrive beyond fall (Phillips et al., 2024). This can lead to the ongoing spread of Lyme disease, which is transmitted by ticks and is prevalent in Georgia and Tennessee. The West Nile virus, common during summer, is notably present in Louisiana and Mississippi. Dengue fever, which is less common in the United States, has been reported in Mississippi.

Increased anxiety and stress: Extreme heat can contribute to higher levels of mental duress (Clayton, 2021). In some instances, prolonged heat can increase irritability. The discomfort associated with heat can also decrease cognitive function.

Application to the Study Area: Northeast Arkansas

Arkansas's rural landscape, characterized by its rich agricultural heritage and tight-knit communities, will experience a multitude of challenges amplified by adverse weather events. With nearly 40% of the state's population residing in these rural areas, understanding the implications of excessive heat is vital for local communities, health professionals, and policymakers (Scott, 2021).

Geographic Context and Challenges

The three primary regions of Arkansas's rural landscape—the Coastal Plains, the Delta, and the Highlands—are distinct with their own ecological and economic characteristics (Miller & Wheeler, 2021). The Northeast Delta region, which includes counties such as Clay, Craighead, Greene, Lawrence, Mississippi, and Randolph, is particularly vulnerable to atmospheric-related changes. Historical data show that average temperatures in this area have risen over the past hundred years, a trend expected to continue in the coming decade (Fan et al., 2024).

Health Implications

The health impacts of elevated temperatures are multifaceted and can be substantial. Elevated temperatures are linked to an increase in heat-related illnesses, such as heat exhaustion and heat strokes, particularly affecting rural communities where farmers spend long hours outside. Furthermore, farm workers often wear protective gear that can limit their ability to sweat and cool down, increasing the risk of dehydration. Prolonged dehydration can have an adverse impact on vital organs, such as the kidneys and liver. Additionally, with their outdoor work environment, farm workers are at greater risk of skin cancer. Beyond physical health issues, the psychological impact of adverse weather events must be considered. Rural areas frequently experience economic instability connected to agricultural viability due to shifting weather patterns. Heightened worries about crop failures and financial struggles can contribute to rising rates of anxiety, depression, and other mental health challenges within these communities.

Agricultural and Economic Impact

Agriculture and forestry are foundational to the rural economy in Arkansas, employing nearly 6% of the state's workforce. As temperatures rise, the implications for agriculture could be substantial. Many staple crops, such as rice, soybeans, and corn, are sensitive to temperature fluctuations. Increased heat may reduce crop yields, impacting food security and the livelihoods of farmers. Furthermore, heat stress can affect the nutritional quality of crops, compounding the issue of food access in rural areas. In addition to crops, livestock health can also suffer from elevated temperatures, leading to decreased fertility, higher disease vulnerability, and increased mortality rates. Farmers may incur additional costs related to cooling methods, veterinary care, and feed, which can strain their financial resources. As agricultural productivity declines, rural economies based on farming may experience downturns, leading to job losses and reduced community investment. This economic decline could result in a reduced tax base, which in turn could impact funding for essential community services, including schools, outreach programs, and infrastructure, further hindering the community's ability to adapt to atmospheric challenges and rural health issues.

Community Support and Quality of Life

As funding for essential services decreases, rural areas may struggle to provide adequate support systems for residents. Programs aimed at enhancing community resilience, offering mental health services, and ensuring healthcare access may not be available, leading to a decline that is difficult to reverse. These ongoing challenges of agricultural viability and economic stability can severely impact the overall quality of life for residents of rural Arkansas. For instance, farmers' markets, fairs, and festivals mark occasions for community bonding and cultural expression in many rural communities. These offerings may now be limited. Finally, the potential loss of jobs may lead to out-migration, particularly among individuals seeking better opportunities elsewhere, which can diminish the vibrancy of these communities.

Application to the Study Area: Southeast Louisiana

The rural landscape of Louisiana, defined by its bayous, swamps, and wetlands, is also expected to contend with challenges stemming from adverse weather events. Consistent with trends observed in other states within the study area, a considerable portion of Louisiana's population resides in rural regions. Furthermore, the rural population has experienced a slight increase over the past decade, rising from 26% to approximately 29%. This demographic shift, combined with the potential health implications associated with extreme heat events, necessitates the attention of business leaders, health practitioners, and the wider community.

Geographic Context and Challenges

Louisiana's physical and geographic landscape can be categorized into three primary regions: the East Gulf Coastal Plain, the Mississippi Alluvial Plain, and the West Gulf Coastal Plain. These regions encompass specific sub-regions, which include North Louisiana (Uplands), Central Louisiana, Acadian, Florida Parishes, and Greater New Orleans. Each region is characterized by distinct ecological and economic attributes that influence their respective vulnerabilities to environmental changes. The Mississippi Alluvial Plain, encompassing parishes such as East Carroll, West Carroll, and Madison in the northern part, as well as East Baton Rouge, West Baton Rouge, and Iberville to the south, will experience challenges stemming from atmospheric changes. Comprehensive analyses of historical climatic data reveal a significant increase in average temperatures within this region over the past century, a trend that is projected to continue in the upcoming decade. This warming raises concerns about the resilience of local ecosystems, the sustainability of agricultural practices, and the overall socio-economic stability of the area, particularly in relation to health outcomes.

Health Implications

In addition to temperature readings, meteorologists and health professionals often refer to the heat index. This measurement provides a more comprehensive assessment of individuals' perceptions regarding the combined effects of heat and humidity. It is particularly relevant in the context of Louisiana, where the average relative humidity is approximately 74%, placing the state among the top three in the nation. Additionally, the average dew point in Louisiana is approximately 60 °F, a metric that some experts suggest may serve as a more accurate indicator of moisture content in the air than relative humidity. Louisiana again ranks among the top three states by this measure, following Hawaii and Florida. Mississippi is ranked fourth while Alabama is fifth. The combination of elevated temperatures, high humidity, and a significant dew point during the summer months contributes to an increased risk of heat exhaustion and heat stroke, particularly among agricultural workers in the Alluvial Plain region. The high humidity levels inhibit the body's natural cooling process through sweating, thereby exacerbating the risk of heat-related health issues (Strelkow et al., 2025). As in other regions, agricultural workers in Louisiana undertake their duties during nighttime hours to accommodate various agricultural practices and enhance operational efficiency. This flexibility in work hours is particularly vital for industries that require extended schedules to maximize productivity. Nighttime operations are especially important for tasks such as irrigation, harvesting, and processing, which may be more effectively conducted during the cooler evening hours to mitigate the adverse effects of high temperatures.

Agricultural and Economic Impact

Agriculture-related sectors constitute an important component of Louisiana's economy, employing approximately 4% of the state's workforce. The onset of summer, characterized by increased temperatures and greater relative humidity, has implications for agricultural practices. Elevated temperatures necessitate higher irrigation expenditures, as farmers in humid environments must supplement soil moisture to sustain crop production. In certain instances, the costs associated with production may surpass the revenue generated, consequently deterring the cultivation of specific crops. This scenario not only leads to economic losses for some agricultural enterprises but also limits the availability of crops during the harvest season. Furthermore, extreme heat conditions may negatively impact the land accessible for forage production. A reduction in forage availability may constrain livestock from meeting their nutritional intake, thereby compelling farmers to incur additional expenditures for feed.

Prolonged heat events disrupt the overall operations of agricultural enterprises. During such periods, agricultural workers are often compelled to extend their work hours, which can result in heightened levels of anxiety and stress. This exposure to demanding conditions increases the risk of heat-related illnesses, such as heat exhaustion, which subsequently elevates the risk of occupational accidents. In some instances, the difficult work environment may function as a contributing factor to the development of depressive symptoms among workers. Collectively, these stressors can adversely impact both the physical and mental well-being of farmers. The cumulative effects of these challenges necessitate a comprehensive understanding of the interplay between atmospheric conditions and agricultural viability in Louisiana.

Community Support and Quality of Life

Humid heat events not only contribute to physical discomfort but also play a role in the dynamics of infectious disease transmission, particularly concerning seasonal influenza. Research suggests that increased humidity levels create a more favorable environment for the survival and spread of pathogens, thereby leading to higher infection rates (Serman et al., 2022). In communities with a substantial proportion of older adults, as well as individuals with chronic conditions such as diabetes and chronic kidney disease, influenza poses a considerable risk and may lead to severe health complications or mortality. In regions experiencing a shortage of healthcare professionals, the healthcare community can enhance individual and collective health outcomes by acknowledging and addressing the effects of humid heat events on the seasonality and transmission dynamics of influenza. This is noteworthy given the aforementioned areas of East Carroll and Madison in Northeast Louisiana and Iberville in Southeast Louisiana.

Additional Challenges

This decade and the next will present challenges for the Southeast United States. Healthcare professionals, policymakers, and anyone invested in enhancing population health and well-being will benefit by actively engaging with these issues. One of the most pressing concerns will be the continued growth of the aging population both nationally and globally (i.e., proportion of the population that is aged 65 years and older). As highlighted in Chap. 4 with the brief introduction of the Epidemiological Transition Theory, older adults are at a greater risk for developing several degenerative conditions, most notably cardiovascular diseases and diabetes. These health issues are disproportionately prevalent in the Southeastern region, exacerbating health disparities and placing additional strain on healthcare systems.

The increased demand for chronic disease management, long-term care, and medical services will necessitate practical approaches in rural areas which currently experience health professional shortages. Due to these trends, integrating elements of geriatric training into the educational framework would benefit health professionals practicing in the six-state study area. This training would provide practitioners with the knowledge required to address the distinct health challenges experienced by the aging population. By focusing on preventive care, early intervention strategies, and the management of chronic conditions, healthcare providers can improve health outcomes for older adults. For example, training in telehealth and telemedicine can facilitate access to care for older adults with mobility challenges or limited transportation options in rural areas. Additionally, training that emphasizes mental health promotion and social connections is practical, particularly since older adults living alone (i.e., widowed, divorced, or never married) often experience reduced social engagement and support. This group of older adults may also lack supportive interpersonal relationships and networks, factors which have been associated with an increased risk of depression.

Another significant challenge facing the Southeast United States is the trend of out-migration from rural areas to urban centers. Although the study area features a higher proportion of rural residents compared to the national average, this demographic group is experiencing a steady decline that has persisted over the past few decades. This out-migration is expected to continue. Several motivations contribute to this urban migration. Economic opportunities often abound in urban areas, providing individuals—especially younger generations—with access to education, employment, and professional advancement that may be limited in rural locales. Additionally, urban regions typically offer a greater variety of services and amenities, such as cultural events, public transportation systems, and social networks, which can be particularly appealing to those seeking a higher quality of life. The implications of this migration are multifaceted. For rural communities, the loss of population, particularly among the young and able-bodied, can lead to a reduced workforce and a decreasing tax base. This decline can further exacerbate existing challenges, such as underfunded education systems, limited healthcare access, and a lack of infrastructural investment. An aging population remaining in these rural areas may experience service shortages, as younger individuals seeking better opportunities fail to return. To address these issues, implementing targeted initiatives that support rural populations while promoting sustainable economic development may be of value. As previously mentioned, investments in high-speed internet and technology can enhance connectivity in rural areas, allowing residents to engage in remote work opportunities and access services online. Encouraging entrepreneurship and small business development in these regions can also create job opportunities that might entice individuals to remain in or return to their hometowns. Additionally, fostering partnerships between urban centers and rural communities can facilitate resource sharing and collaborative efforts to improve quality of life. By embracing innovative solutions that enhance the viability of rural living while addressing the challenges posed by current migration patterns, stakeholders can work towards a population distribution that benefits all regions of the Southeast United States.

Population Health: Policies

While I addressed various health policies that promote population health in Chap. 3, I dedicate the remainder of this chapter to emphasizing key initiatives that offer the potential to further reduce health inequities. One critical area of focus is the ongoing effort to reduce the number of rural counties classified as health professional shortage areas. Ensuring that healthcare professionals are available in these underserved regions is vital not only for providing direct medical care but also for serving as a reliable source of health information. Effective management of chronic diseases necessitates a comprehensive approach that incorporates not just treatment but also timely information, emotional support, and encouragement from knowledgeable professionals. Therefore, we must support programs that incentivize healthcare practitioners to relocate to and work in rural areas (Dolea et al., 2010; WHO, 2021). Policymakers should explore innovative incentive structures, such as student loan forgiveness, housing allowances, or professional development opportunities, to attract and retain more healthcare professionals in these regions.

Second, business leaders and community organizations can work collaboratively to forge partnerships with local churches, libraries, schools, and community centers to host regular health education events. Such gatherings can serve as supportive environments that not only disseminate health information but also promote positive health behaviors among community members (Azul et al., 2021). By creating welcoming spaces for health-focused discussions and activities, we can foster a culture of wellness that encourages residents to engage in healthier lifestyle choices. These collaborations can also strengthen community ties by providing a network of support for those striving to improve their health outcomes. Local policymakers can actively contribute by supporting these events, presenting awards, and highlighting the importance of civic engagement within the community. Their involvement also fosters a culture of participation and empowerment among citizens.

Finally, individual effort is instrumental for improving population health and well-being. Each person's commitment to adopting healthier lifestyle choices enhances their personal well-being and positively impacts the collective wellness of the community. By prioritizing nutritious diets, regular exercise, and mental wellness, individuals can inspire those around them—family, friends, and colleagues—to adopt similar habits. Additionally, individual contributions, such as volunteering for health initiatives, supporting local health policies, and advocating for better access to healthcare, can improve community health outcomes. Engaging in preventive care, participating in health screenings, and adhering to medical guidance can lead to early detection of diseases and reduce healthcare costs. Personal responsibility in managing one's own health can alleviate some of the burdens on rural public health systems, allowing resources to be directed toward those who need them most.

Concluding Remarks

I would like to conclude by sharing additional resources for readers interested in learning more about rural health issues in the Southeast United States.

A key resource is the American Community Survey (ACS), which is conducted annually by the Census Bureau. This survey provides valuable demographic, economic, and some health-related data across various geographic levels (e.g., state, county). More information can be found at https://www.census.gov/programs-surveys/acs/

For insights into health facilities, healthcare professions, hospitals, and population characteristics, the Area Health Resource Files (AHRF) is a useful resource. Managed by the Health Resources and Services Administration (HRSA), this dataset can be accessed at https://data.hrsa.gov

Another valuable resource is the Behavioral Risk Factor Surveillance System (BRFSS), which gathers information on health behaviors and conditions from residents nationwide. Interested readers can learn more by visiting this site: https://www.cdc.gov/brfss/index.html

An important aspect of this book was the special chapter on emerging health issues. To better understand and assist communities, readers should explore the Community Resilience Estimates (CRE) website. Developed by the Census Bureau, the CRE helps identify vulnerable areas in the face of disasters. Both sites are worth exploring, especially the interactive features of the CRE: Census at https://www.census.gov/programs-surveys/community-resilience-estimates.html and interactive CRE (see CRE Interactive Tool), as well as data sources.

The final resource that I highlight is the Diabetes Surveillance System. This interactive application provides extensive information on diabetes prevalence and risk factors. Additionally, the social determinants of health application feature national and state-level maps highlighting diagnosed diabetes prevalence based on various social determinants. The website link is https://gis.cdc.gov/grasp/diabetes/DiabetesAtlas.html

In conclusion, this book serves as an invaluable resource for understanding and addressing the rural health needs of the Southeastern United States. Drawing on my experience in health economics, population health, public policy, and spatial analysis, I provide a comprehensive examination of the unique health challenges faced by rural communities in Alabama, Arkansas, Georgia, Mississippi, Louisiana, and Tennessee. These regions grapple with several health issues exacerbated by poverty, outdated infrastructure, and limited healthcare access.

By employing an interdisciplinary approach, the book integrates principles from economics, geography, history, and sociology, alongside the latest research on mortality and health outcomes. The narrative begins with definitions of rural and health and then progresses through a historical context, a discussion of social and racial considerations for health, and an exploration of key health conditions such as coronary heart disease, diabetes, and obesity. Next, it addresses emerging health threats linked to natural disasters, emphasizing the urgent need for resilient healthcare infrastructures. The place-based examples focusing on communities in Arkansas and Louisiana also offer practical insights. The book employs a structured approach by emphasizing key areas related to the leading causes of mortality and emerging health issues, thereby offering a concise exploration of health outcomes in the study area. While recognizing the importance of additional health-related topics, such as health literacy, mental health, and pandemics, I believe these subjects are better suited for examination by scholars with specialized expertise in their respective fields. This specialization ensures a more informed analysis, thereby contributing to the advancement of knowledge and understanding within these critical areas of public health.

The book not only summarizes critical health issues but also offers a wealth of resources for further exploration. It is designed for healthcare professionals, researchers, policymakers, and anyone committed to improving health outcomes in these underserved regions. By bringing awareness and facilitating informed discussion, this work aims to inspire collaborative efforts toward meaningful solutions for rural health challenges in the Southeast United States.

References

Azul, A. M., Almendra, R., Quatorze, M., Loureiro, A., Reis, F., Tavares, R., & Ramalho-Santos, J. (2021). Unhealthy lifestyles, environment, Well-being and health capability in rural neighbourhoods: A community-based cross-sectional study. *BMC Public Health, 21*, 1–18.

Beard, C. B., Eisen, R. J., Barker, C. M., Garofalo, J. F., Hahn, M., Hayden, M., & Schramm, P. J. (2016). *Vector-borne diseases* (pp. 129–156). US Global Change Research Program.

Clayton, S. (2021). Climate change and mental health. *Current Environmental Health Reports, 8*, 1–6.

Dolea, C., Stormont, L., & Braichet, J. M. (2010). Evaluated strategies to increase attraction and retention of health workers in remote and rural areas. *Bulletin of the World Health Organization, 88*(5), 379–385.

Fan, X., Miao, C., Wu, Y., Mishra, V., & Chai, Y. (2024). Comparative assessment of dry-and humid-heat extremes in a warming climate: Frequency, intensity, and seasonal timing. *Weather and Climate Extremes, 45*, 100698.

Heidari, H., Mohammadbeigi, A., Khazaei, S., Soltanzadeh, A., Asgarian, A., & Saghafipour, A. (2020). The effects of climatic and environmental factors on heat-related illnesses: A systematic review from 2000 to 2020. *Urban Climate, 34*, 100720.

Ingram, K. T., Dow, K., Carter, L., Anderson, J., & Sommer, E. K. (Eds.). (2013). *Climate of the Southeast United States: Variability, change, impacts, and vulnerability*. Island Press/Center for Resource Economics.

Kuehn, B. M. (2023). Initiatives aim to help mitigate cardiovascular risks of extreme temperatures. *Circulation, 148*(25), 2064–2066.

Masson-Delmotte, V., Zhai, P., Pörtner, H. O., Roberts, D., Skea, J., Shukla, P. R., & Waterfield, T. (2019). Global warming of 1.5 C. In *An IPCC special report on the impacts of global warming of 1* (pp. 93–174).

Miller, W., & Wheeler, E. (2021). *Rural profile of Arkansas Social & Economic Trends Affecting Rural Arkansas*. University of Arkansas, Division of Agriculture Research and Extension.

Phillips, M. C., LaRocque, R. C., & Thompson, G. R. (2024). Infectious diseases in a changing climate. *JAMA, 331*, 1318.

Scott, R. (2021). *Rural profile of Arkansas, 2021*. University of Arkansas, Division of Agriculture. https://www.uaex.uada.edu/publications/pdf/MP564.pdf

Serman, E., Thrastarson, H. T., Franklin, M., & Teixeira, J. (2022). Spatial variation in humidity and the onset of seasonal influenza across the contiguous United States. *GeoHealth, 6*(2), e2021GH000469.

Strelkow, A., Mitchell, T. J., Knapp, P. A., & Ortegren, J. T. (2025). Assessing the frequency, duration, and spatial extent of summertime extreme dew point conditions in the southeastern USA, 1973–2022. *Atmosphere, 16*(3), 265.

U.S. Environmental Protection Agency, EPA. (2024). *Climate change indicators: Weather and climate*. EPA. https://www.epa.gov/climate-indicators/weather-climate

Vaidyanathan, A. (2024). Heat-related emergency department visits—United States, may–September 2023. *MMWR. Morbidity and Mortality Weekly Report, 73*, 324.

World Health Organization, WHO. (2021). *WHO guideline on health workforce development, attraction, recruitment and retention in rural and remote areas*. World Health Organization.

Index

A. Sharma, *Current and Emerging Rural Health Issues in the Southeast United States*, Global Perspectives on Health Geography,
https://doi.org/10.1007/978-3-032-07543-7

MIX
Papier aus verantwortungsvollen Quellen
Paper from responsible sources
FSC® C105338

If you have any concerns about our products,
you can contact us on
ProductSafety@springernature.com

In case Publisher is established outside the EU,
the EU authorized representative is:
Springer Nature Customer Service Center GmbH
Europaplatz 3, 69115 Heidelberg, Germany

Printed by Libri Plureos GmbH
in Hamburg, Germany